Criminal Justice
Recent Scholarship

Edited by
Nicholas P. Lovrich

A Series from LFB Scholarly

Accountability by Camera

Online Video's Effects on Police-Civilian Interactions

Douglas A. Kelly

LFB Scholarly Publishing LLC
El Paso 2014

Library of Congress Cataloging-in-Publication Data

Library of Congress Cataloging-in-Publication Data

Kelly, Douglas A., 1963-
Accountability by camera : online video's effects on police-civilian interactions / Douglas A. Kelly.
pages cm. -- (Criminal justice: recent scholarship)
Includes bibliographical references and index.
ISBN 978-1-59332-757-6 (hardcover : alk. paper)
1. Police-community relations. 2. Imaging systems. 3. Video recordings. I. Title.
HV7936.P8K47 2014
363.2'3--dc23

2014011591

ISBN 978-1-59332-757-6

Manufactured in the United States of America.

DEDICATION

This book is dedicated to my spouse.

Table of Contents

Acknowledgments

The author would like to acknowledge the assistance and guidance of Michael Socolow, Paul Grosswiler, Howard Segal, Nathan Godfried, and Jeffrey St. John.

CHAPTER 1

Introduction

Video captured by increasingly ubiquitous citizen cameras and communicated to a mass audience over the Internet is capable of bypassing police jurisdictional influence over traditional mass media and may be affecting police-civilian interactions in American public space as the initial cusp of a paradigm shift.

The growing number and wide distribution of cameras viewing public spaces in America is reaching levels unprecedented in our history. In part, this represents a return to earlier eras in which it was difficult for an individual to go unwatched or unidentified in their own community. However, the camera is inherently different from the eyewitness in that it produces an *evidentiary record.* Historically, the ability to visually record activities in public space was reserved to those with the resources and the motivation to devote considerable finances, time, and technical expertise to the task. Today those necessary resources have shrunk to be within reach of the vast majority of Americans. Deploying cameras to record public space is no longer a question of feasibility, but simply one of choice.

Historically, both the police and the traditional mass media wielded considerable power through cameras, power that has not been available to the general public. Autonomous citizens working for social change were usually at a significant disadvantage, and were often effective only when pooling resources to levels approximating the larger organizations they opposed. Today, police seeking to image public space often find their cameras outnumbered by those under autonomous citizen control. Further, police imagery has sometimes been used counter to its original intended purposes after release in compliance with state-level public records "sunshine laws."

Internet tools for mass communication of video, such as Flickr and YouTube, in combination with the new ubiquity of high resolution cameras, are empowering the autonomous citizen in American public space in ways denoting the initial cusp of a paradigm shift. A button press on a relatively cheap cell phone can instantly publish user-generated video to Internet servers available to a world audience, beyond the police jurisdiction of the space being imaged. Arguably, police leverage on local media outlets is no longer sufficient; local control of imagery appears to have been lost.

Police and citizen attitudes toward the new ubiquity of imaging systems are mixed; historically, organizations and individuals have often acted to protect their own privacy, but have also often acted to preserve their ability to image others. This overall pattern has not changed since the invention of instantaneous photography at the end of the 19th century.

Police-civilian public space interactions are the lowest level, highest stakes interactions in the United States. Police are granted state powers up to and including lethal force. Police have discretionary powers to immediately restrict or take a civilian's life, liberty, or property. While such actions may be reviewed at a later time and place, the possibility of such actions can have a chilling effect on civilian action in public space.

Police powers are restricted by a system of which police officers are a major part. The courts, including judges, prosecutors, and police are "colleagues" in many respects. For a variety of reasons explicated repeatedly in the research literature over the last century and a half, police testimony tends to be accepted over that of a civilian, both by the court officials directing the proceedings and by juries. Part of this deference to the officer's testimony can be attributed to the traditional respect accorded the police, and part of it represents an acknowledgment of police training and standards. However, a significant part of the reluctance to question police testimony has been attributed, by commissions, jurists, scholars, and the police themselves, to the unwillingness of the court system to contradict and thus challenge the police with whom the courts must work on a continuing basis.

"*Quis custodiet ipsos custodes?*" [Who will guard the guards themselves?] (Juvenal, *Satires*, 6.347-48, ca. 1st century C.E.). It has been explicated in the literature that police will sometimes lie to protect themselves and other police, including committing perjury, falsifying or destroying evidence, and making false reports (Klockars, et al., 2007). Estimates of the frequency of this behavior range from a significant minor fraction to a majority fraction, as reported by commissions, jurists, scholars, and the police themselves (Loevy, 2010). This is apparently an organizational behavior pattern, not the result of "a few bad apples," despite the common offering of that phrase as a defense. The literature also appears to document this behavior consistently across time and geography. In particular, as regards the research carried out for this book, it has been documented that police have, on occasion, destroyed or denied the existence of video evidence of police misconduct, when such evidence was within their exclusive control (Balko, 2011).

Reports also exist documenting the fact that police officers in the field on occasion have acted in contravention of court rulings and their own departments' operating orders in their handling of photography of public spaces (New York Police Department, 2009); more specifically, police and prosecutors have sometimes misused wiretap laws to harass and prosecute videographers in violation of the First Amendment (Balko, 2011). On the other hand, reports also exist indicating that some Internet videos have been edited to show police officers in the worst possible light (Halbfinger, 2009). Reports exist that, on occasion, a media outlet has refused to distribute videos of apparent local police misconduct (Tompkins, 2010); on the other hand, reports also exist showing that some incidents have been staged and selectively edited to produce such videos (Halbfinger, 2009).

Based on historical trends, it is likely that the technological developments underlying these problems are not going to go away or to decrease; rather, they will tend to increase and to make police work ever more challenging.

The stated problems have significant implications for the continued exercise of First Amendment rights in photographing public space, both for autonomous citizens and for professional journalists. The stated problems also have significant implications for daily civilian life in the United States, particularly with respect to police accountability and Fourth Amendment protections against search and seizure and arrest without probable cause.

At the time of this writing, there appears to have been minimal scholarly research on this contemporary phenomenon. It is hoped that the products of the present research will reveal opportunities for further scholarly research in several fields.

The scope of the present research is limited in time to the period from 2005 (YouTube went online in the summer of 2005) through 2010 (for the date of the initial police-civilian interaction), and in space to the United States of America - that is, the geographical area under U.S. law, including states, commonwealths, districts, and possessions. These limits usefully restrict the research to a single legal system, a single legal language, and a political system with Constitutional traditions of free speech protection and open government.

Public space for the purposes of this research is as legally defined according to U.S. Supreme Court rulings. Although such rulings have varied somewhat historically, for the research time period the rulings are relatively consistent. Public space has been consistently held by the courts to include streets, sidewalks, parks, and other public property. Protected speech activity (including photography) in public space that is historically a public forum is most strongly protected, followed by public space that is not a traditional public forum.

The restriction to public space is useful to the present research because: 1) police are not legally empowered to arbitrarily prevent citizens from using a public space for protected speech activities; and, 2) there is a significant reduction in the reasonable expectation of privacy in a public space. Consequently, the case studies examined are less legally complex or ambiguous as to whether the civilians had a right to be present or to record video of the police and other subjects.

Historically and recently, the state has extended police powers to a broad range of persons. Thus, the term ‘police’ as used in this research includes any law enforcement officer, private security guard, or other person granted police powers to act for a government agency while interacting with civilians in American public space.

CHAPTER 2

Context and Precedents

At the present time, there is no one resource that impartially and rigorously gathers and examines the necessary data from the various disciplines touching on these issues. However, many of the issues have been examined individually within the confines of a single discipline, with results that can be usefully synthesized to provide context. A review of the issues as examined in the most pertinent disciplines clearly delineates a void in the literature that is productive for original research.

The first part of this chapter is a brief history of American police-civilian interactions, with particular attention to patterns of police accountability and police responses to civilian photography prior to 2005. This chapter also incorporates a brief legal history of public space imaging in America, with particular emphasis on the present-day legal environment surrounding these issues. The final pieces of this chapter survey the history of imaging technology, examining what scholars have written about the technological development of cameras, their costs, what American social and individual attitudes have been towards them, and their technical capabilities and limitations up to the present day. This is a historiography for most of the period examined, with the most recent data drawn from primary sources.

The void in the literature is: the influence (if any) of user-generated online video on police-civilian interactions in American public space, from 2005 to the present.

POLICE-CIVILIAN INTERACTION AND PHOTOGRAPHY

> Every policeman has to exercise personal discretion in his duties - decisions about when and how to act, whom to suspect and whom to arrest. Such choices are the most important part of his work, distinguishing the policeman from the soldier who does not act without direct orders. (Miller, 1975, p. 85)

The insular nature of police culture, and the (to date) effective actions of police organizations to preserve that insular culture, have had the effect of separating police from the heterogeneous American mainstream culture. This is evident each time an internal police review of a questionable police-civilian interaction renders a decision that the general population, the media, and the civil authorities publicly characterize as a 'slap on the wrist.'

Any increase in visibility, such as badges, uniforms, and badge numbers, necessarily decreases the ability of police to exercise discretion in the maintenance of public order. Vila & Morris (1999) cite a number of scholars in asserting that, if police can be individually identified and linked to their interactions with civilians, police can more readily be held accountable for abuses of authority (pp. 35-39). Prior to photography, such disputes were resolved on the relative weights of civilian versus police testimony. Photographs present only a single moment in time; police testimony could still present a viable interpretation at odds with civilian testimony. However, motion pictures present a sequence of events that are much more difficult to explain away convincingly. Finally, multiple simultaneous motion pictures, from independent cameras with different angles of view, are practically incontrovertible.

The 60-pound tripod-mounted camera of the mid-1800s was obtrusive enough to give police ample warning of its presence. Concealable 'detective' cameras of the 1880s made it possible to photograph police candidly. Fast shutters and high-speed film enabled capturing motion clearly in available light without an obtrusive flash. Telescopic lenses enabled imaging far beyond the reach of a beat cop's arm. However, the physical medium of film meant that a camera was susceptible to being confiscated or broken, the film exposed, and evidence thereby destroyed. Similar actions are possible for any camera using a recording medium. Mass duplication and public distribution depended on newspapers, theaters, and television stations, many of which had close relationships with local police. Thus, police retained the ability to exercise discretion in the maintenance of public order.

Since 2005, common cell phone cameras have been capable of capturing still images and motion pictures and instantly sending that evidence to image and video sharing services on the Internet, thereby placing the evidence beyond the immediate reach of local police and within easy view of hundreds of millions of Internet users. This marked increase in visibility is likely to decrease significantly the ability of police to exercise discretion in the maintenance of public order. The ability of civilian videographers to incontrovertibly identify and link police to their interactions with civilians may mean that police can more readily be held accountable for abuses of authority. This informs the research question: *What is the outcome of user-generated online video on police-civilian interactions in American public space?*

Police responses have shown an increasing awareness of the connection between civilian photography of police-civilian interactions and challenges to police exercise of personal discretion. Police responses have included confiscation of cell phones or cameras, coerced destruction of images, and arrest or threat of arrest, often in direct contravention of law, departmental policies, or judicial orders. Historical patterns of police organizational efforts to protect police culture seem to indicate that these responses will likely continue.

Barkan & Bryjak (2011) provide useful sociological analyses of police culture and a finding aid to more narrowly focused scholarship. Their chapter on police misconduct is particularly valuable to the present research. The authors elucidated one reason for the phenomenon of police who used excessive force against civilians who defied or criticized them: "Some officers consider a verbal affront to be no different than a physical assault, inasmuch as verbal defiance can lead to a loss of control, which in turn increases the risk of danger" (p. 295). It is therefore not unreasonable for police to resort to force in an attempt to regain control of the situation (Barkan & Bryjak, 2011, p. 295). This observation is crucial to understanding a common conflict between police and civilian cultures, and which appears in a number of the case studies in the present research.

Barkan & Bryjak (2011) also argue that, historically, police internal affairs investigations have not met with much success. They observe that IAD investigations were rife with conflicts of interest, were not perceived as legitimate by the police rank and file, and that high-ranking police saw them as threats to their jobs and to a good public image of the department (p. 297). This argument is central to the present research, as the majority of the case studies include at least one police internal affairs investigation.

The authors' examination of community policing (CP) is valuable to the present research because it has a number of important parallels to the possible outcomes of user-generated online video of police-civilian interactions. CP holds promise for improving accountability for both police and civilians, but Barkan & Bryjak (2011) note that it requires the "blue wall" to become transparent. For police to support that transparency, they need to trust the community partners to look out for police interests as well as their own.

Vila & Morris (1999) have edited a highly useful documentary history, supplemented by a number of insightful and well-reasoned essays that are of value to the present research. The perspective of Vila, formerly a police officer for 17 years, informs and validates the evident scholarship of this work.

Vila & Morris (1999) point to the Lexow Commission, created in 1894 to investigate the New York City police at the direction of the state legislature, as possibly the most prominent example of early police reform efforts. The ensuing exposure of corruption led to the weakening of the Tammany Hall political machine and the brief but significant administration of Theodore Roosevelt as President of the Board of Police Commissioners, 1895-7 (Vila & Morris, 1999, p. 62-67). This speaks directly to the present research, in providing historical context for the resistance of police departments to oversight by elected officials.

As noted previously, Vila & Morris (1999) cite a number of scholars in asserting that if police can be individually identified and linked to their interactions with civilians police can more readily be held accountable for abuses of authority (pp. 35-39). This is a significant assertion for the present research in regard to issues of accountability involving new technologies for individually identifying police and linking them to specific police-civilian interactions.

The Rodney King beating is one of the cases that Vila & Morris (1999) choose to examine in substantial detail (pp. 266-267). This is particularly relevant to the present research for its high-profile introduction of civilian video into the public debate over police accountability. Vila & Morris (1999) note that, in the aftermath of the King incident, several experts testified in favor of community policing as a solution to many of the LAPD's problems (p. 269).

In a 1992 essay, Vila suggests that the persistent problems of the police 'code of silence' and institutional failures to prevent improper use of force are not due to selection problems, but to society's unrealistic expectations. He points out that "institutional cultures that portray police as crusaders against crime and the unrealistic physical and emotional demands of the job" (p. 273) put police in situations where they will inevitably make illegal yet understandable mistakes. To avoid the severe penalties, police cover up these mistakes for one another. "Once an officer has been 'covered for' by peers, he or she owes them the same favor in return" (Vila & Morris, p. 273). This is a concise definition and explanation of a phenomenon that is central to the present research.

In another selection by Vila & Morris (1999), Klockars (1996) wrote that three major obstacles to controlling excessive police use of force all come from "the fundamentally punitive orientation of the quasi-military administrative apparatus of American police agencies" (p. 282). The three obstacles are "the code," the CYA or "cover your ass" syndrome, and the attitude among line officers and supervisors that "the 'good' supervisor is the one who will 'back up' an officer when he or she makes a mistake" (Vila & Morris, 1999, p. 282). Klockars' obstacles are highly useful in contextualizing and characterizing the data in many of the present research's case studies.

In a later book with several co-authors, Klockars et al. (2007) presents the results of survey and case study research of over 3,000 police officers in 30 law enforcement agencies, which was originally funded by and reported to the National Institute of Justice in 2001. This work is relevant to the present research because the authors reported that police car cameras eliminated opportunities for some types of police misconduct:

> Because all traffic stops in Charlotte-Mecklenburg are videotaped, this changed the entire meaning of the incidents we presented to them. It vastly reduced the opportunities for, and increased the consequences of, misrepresenting or failing to report the full details of an incident of the type we described actually happening. Repeatedly, our focus groups told us that

"this kind of thing" just wouldn't happen in their department. (Klockars et al., 2007, p. 241)

This finding directly addresses the issue of accountability as it is examined in the present research.

In Klockars et al.'s (2007) study, one of the traffic stop scenarios used to discuss the police code of silence was called "Arrest an asshole day" (pp. 239-249) in which a rude and verbally abusive driver was replied to, in kind, by a police officer while a fellow officer witnessed the exchange. While most police emphasized the importance of immediately separating the driver and the angry officer, opinions differed on whether to report the incident. Significantly, this was the scenario in which police who worked with in-car video recording stated that that sort of exchange simply didn't happen. The responses from police in other organizations that did not use in-car cameras varied slightly, but generally agreed that the driver was going to complain, the citation should be issued, and that supervisors should be informed: "Supervisors don't like to be surprised" (Klockars, et al., 2007, p. 242). The authors noted that most police advocated filing the internal report not because it was the right or ethical thing to do, or even that it was standard operating procedure, but because discourtesy to a civilian carried a suspension of a day or two whereas lying on internal paperwork was a firing offense. As one detective put it, "The only time worth lying is to cover up something you've done that's so serious you'll get fired for it anyway" (Klockars, et al., 2007, p. 247). This finding is directly applicable to several of the case studies in the present research.

Continuing the theme of Klockar's traffic stop scenario, Crank (1998) argues that the exercise of discretion by police may rest on cues so subtle as to appear intuitive. "Instinct, as much as the presence of articulable cause, may guide their behavior. And arrest may represent a desire to bring justice to some asshole rather than a calculated estimate of the presence of probable cause" (p. 34). This explication of police behavior is useful in contextualizing several of the police-civilian interactions in case studies in the present research.

Crank (1998) also argues that police behaviors are not simple, but stem from several influences: "Contact with the public is articulated through the organization. The decision to intervene is at the discretion of the officer, but problems associated with intervention are enveloped in department policy" (p. 31). Furthermore, the author observes the strongly negative influence of the core guiding document in most police organizations, the "Standard Operating Procedure is a typically thick manual that defines the vast array of rules telling officers what they should not do in various circumstances, representing, quipped one officer, '100 years of fuckups'" (Crank, 1998, p. 33).

In an argument that is particularly relevant to the issue of cameras and police accountability, Crank (1998) states that news media personnel are regarded by police as members of the civilian public, but with special powers, and are therefore accorded special treatment. "Reporters are social control agents whose influence can both negatively and positively affect the police. The police know this, and seek their sanction through co-optive strategies" (Crank, 1998, p. 35). This observation implies one of the challenges for police in the present research due to the increasing number of civilian cameras that are not mediated through a co-optable news organization.

Monkkonen (1992) argues that the adoption of a modern police force generally followed the growth of the urban population; New York, Boston, and other fast-growing cities were the earliest centers to make the change, following a diffusion curve common to a number of innovations (p. 553). This pattern remains apparent even in the present research, where the majority of the studied police-civilian interactions take place in the New York, Boston, and Baltimore metropolitan areas.

Monkkonen (1992) identifies four innovative features of nineteenth century American police: hierarchical paramilitary organization (modeled after Sir Robert Peel's London Police of 1829), functional differentiation from the judicial to the executive branch, visible uniforms, and expectations of activity (pp. 550-551). American police culture developed more strongly in the direction of maintaining public order rather than exclusively enforcing laws for the courts. This change in emphasis meant that police on the street were expected to exercise more discretion in what actions were necessary to maintaining order. This difference in American policing was a factor in defining the scope of the present research, and may argue for the limited applicability of the research findings to jurisdictions with differing police traditions.

Monkkonen (1992) also notes that the per capita number of police in America grew from 1860 levels of around 1.3 per thousand to 2 per thousand in 1908 (p. 554). West (2000) observed that by 1910, roughly one-third of the U.S. population owned a camera (pp. 74-75), largely due to Kodak Brownie sales. This meant a ratio of more than 150 cameras for each police officer in the United States, a significant point for the historical context of the present research.

Miller (1975), as quoted at the beginning of this section, makes a number of arguments about the first 40 years of professional New York City police that are useful to an understanding of present-day police practices. According to Miller (1975), nineteenth century New York police sometimes charged people with disorderly conduct when the actual behavior was disrespect for the officer's authority (pp. 87-88). In addition, the author notes that "Low convictions in proportion to arrests can make policemen into frustrated antagonists of the judiciary, ready to substitute street-corner justice for procedural regularity" (Miller, 1975, p. 89). These points provide historical context for similar arrests in several of the case studies in the present research.

Miller (1975) argues that the New York police officer's personal authority was often based less on concern for legal restraints and more on the unregulated discretion of the individual officer. By tolerating this, officials demonstrated that they trusted men but distrusted institutions (p. 90). The conflict between traditional police practices of personal authority and the exposure of those practices through user-generated online video is central to the present research.

Social historian Fogelson (1977) examines police reform efforts with a primary emphasis on the administration and politics of large metropolitan police departments, but with enough attention paid to street-level policing that this work provides context that is useful to an understanding of events in several of the case studies in the present research. His thesis is that reformers of American police organizations have only had mixed success in reaching their goals, but have seriously exacerbated problems in police-community relations.

Fogelson (1977) argues that one of the most significant findings of the 1894 Lexow Committee was that police "assaulted ordinary citizens with impunity" and "frequently intimidated, harassed, and otherwise oppressed the defenseless and law-abiding citizens whose protection was their central duty" (pp. 3-4). These quotes would not be out of place in a number of the case studies in the present research, and thus provide evidence that these forms of police misconduct have a long history.

According to Fogelson (1977), the period between the police reform efforts of the 1890s and the 1930s saw a significant change in police culture at the street level, but that change was an unintended consequence of the reform efforts; most police felt isolated and alienated from the communities they served. Fogelson (1977) argues that, under sustained attack from reformers and society in general, the police perceived themselves to be at odds with civilian society, a form of *occupational paranoia* that in due course became an integral part of police occupational identity (pp. 110-116).

Fogelson (1977) notes that positive long-term reform effects on street police were not easy to find. One result was that, according to public opinion polls, police ranked in prestige above janitors but below stenographers (Fogelson, 1977, pp. 102-103). The issue of top-down police reform meeting resistance at street level is apparent in more than one of the case studies in the present research.

According to Fogelson (1977), testimony before the Lexow Committee and other evidence seems to indicate that the Keystone Kops' antics bore more than a passing resemblance to actual police behavior. Nevertheless, such portrayals annoyed the International Association of the Chiefs of Police so much that in 1913 they passed a resolution at their annual meeting to do as much as possible to change those depictions (Fogelson, 1977, pp. 112-13). This stated awareness by police chiefs of the public relations value of motion pictures speaks directly to the historical context of the user-generated online videos at the heart of the present research.

Taking a social behaviorist perspective, Sykes & Brent (1983) argue that a situation that ends in violence or an arrest is often a police failure: "A measure of a good police officer is his ability to handle a difficult situation without use of violence, and, in the case of minor violations, without arrest" (Sykes & Brent, 1983, p. 25). This argument may be applicable to a number of the case studies in the present research, particularly those where prosecutors or the court dismissed the initial charges against the civilian subject.

According to Sykes & Brent (1983), one of the challenges to the professional model of policing, and one of the advantages cited by proponents of community policing, is the difference between the definitions of "order" in the cultural backgrounds of nonresident police and in the culture(s) of the community.

> Police have been socialized into a multitude of private social orders. They identify their private order with ORDER. Only in time do some learn that many such orders exist, and they learn to "live and let live." In the meantime, they may proactively intervene because others do not observe their own customs. (Sykes & Brent, 1983, pp. 28-29)

This challenge may be evident in one or more of the case studies in the present research, particularly those where the police and the civilian subject or videographer come from divergent cultural backgrounds.

The authors argue that, between police and civilians, "the "wrong" set of acts by one may bring out what appears to be an uncharacteristic response of the other. In reality it is not uncharacteristic at all, merely the outcome of an unusual transaction" (Sykes & Brent, 1983, p. 253). This may be a significant consideration for the present research; it is possible that a civilian pointing a camera at police may elicit a strong police response simply because the civilian's action is unusual.

Law professor Skolnick (1982, 1994) has written extensively on street-level policing, and provides three well-supported arguments that are crucial for the present research. First, the author argues that discretion remains a core attribute of street-level policing: "Cops will always exercise low-level discretion – influenced both by the culture of policing and the pressures and understandings of the organizations within which they work" (Skolnick, 1994, p. x).

Second, Skolnick (1982) argues that "The law often, but not always, supports police deception" (para. 6). He elaborates that perjury in the courtroom is accepted by police culture because "The end justifies the means" (Skolnick, 1982, para. 9).

Finally, Skolnick (1994) cites a number of scholars who, examining civil rights actions and commission reports, have concluded that police-civilian interactions that do not reach a courtroom are not protected from police misconduct by any of the supposed deterrent effects of the exclusionary rule. Skolnick (1994) argues that police simply work around this rule (pp. 205-212).

Loevy (2010), a practicing attorney, attests to his own experience of the phenomenon of police "testilying" as evidenced by a change in instructions to Chicago juries over fifteen years. The author writes that in the mid-1990s, plaintiffs in police abuse cases would ask that jurors be instructed "to counteract a prevailing assumption that the police were always right and should be believed" (para. 6). In 2010, he writes that defense counsel for the police asked for jury instructions in nearly the same language, but intended to "remind juries that the testimony of police officers deserves their fair consideration as well and that such testimony should not be automatically discounted just because they are police" (Loevy, 2010, para. 7). Further, Loevy (2010) cites Judge Alex Kozinski, chief judge of the Ninth Circuit, interviewed in *The American Lawyer*, "It is an open secret long shared by prosecutors, defense lawyers and judges that perjury is widespread among law enforcement officers" (Loevy, 2010, para. 16). Finally, the author argues that "Even when evidence unequivocally proves falsification (say in a videotape), judges are loath to brand an officer a falsifier" (Loevy, 2010, para. 30). This last argument is particularly relevant to the several case studies in the present research where police testimony is contradicted by user-generated online video.

Jacobi (2000) is one of the previously mentioned scholars who have examined civil rights actions and commission reports, and concluded that police-civilian interactions that do not reach a courtroom are not protected from police misconduct by any of the supposed deterrent effects of the exclusionary rule (Jacobi, 2000, pp. 806-811). This conclusion is important to the present research, because a number of the case studies presented evidence of further cases, lacking user-generated online video, where police misconduct never reached a courtroom. The author also argues that the failure to prosecute police has been blamed on the conflict of interest between prosecutors and police, who are necessary for the prosecutors to bring in convictions (Jacobi, 2000, pp. 802-11). This issue is apparent in several of the case studies in the present research, in which prosecutors continued to support police despite the contrary evidence of user-generated online video.

Quirke (2008) reexamines an earlier example of police control of motion pictures of police actions - in particular, the 1937 Memorial Day Massacre at the Republic Steel plant in South Chicago in which 300 police shot and beat a crowd of strikers and their supporters who were organizing a mass picket. Police bullets struck forty civilians, killing four at the scene; another six died later. Quirke's (2008) reframing presents an invaluable perspective on this controversial incident and its sequelae. She notes that the first published reports edited the newsreel footage and still photographs to tell a pro-police, anti-union story, "purporting to show strikers as a 'riotous mob'" (Quirke, 2008, p. 130). The story from labor's point of view only emerged from the intervention of Paul Anderson of the *St. Louis Post-Dispatch*, and the pro-labor Senators Elmer Thomas and Robert La Follette of the La Follette Committee. Quirke argues that Anderson and La Follette attained a rereading of the images that placed responsibility for the Massacre on the police instead of the strikers (Quirke, 2008, p. 145). "Many times during their hearings, Senators La Follette and Thomas assured witnesses that the police were not on trial. When the committee's work was finished, however, the police stood condemned" (Quirke, 2008, p. 147). It is unlikely that any Chicago officer would ever again grin conspiratorially at a movie camera, as one did in the Lippert footage that Paramount suppressed (Quirke, 2008, pp. 134, 144). These events, and Quirke's analysis, present useful historical context for a number of issues relevant to the present research, particularly self-censorship of media outlets in cooperation with police, and the power of motion pictures to expose police misconduct.

The innovation of wire photo services in the late 1930s, as presented by Hannigan & Johnston (2004), and the effects of that innovation on police accountability, include a number of useful parallels to the present research. The evident level of competition between the news photo services is also significant to the present research, because it presents comparable circumstances for distribution of images of police misconduct; with so many avenues of distribution, it becomes more difficult for police to ensure the suppression of imagery.

> What the wire services did was create the potential for an event to be changed from a personal or regional experience to a national cultural experience. The image now operated as a link, unifying the American culture through this shared experience. (Hannigan & Johnston, 2004, p. 9)

The relevance of this passage to the present research is clear; an argument could be made that substituting 'user-generated online video' for 'wire service' presents an equally valid statement.

Carlebach (1992) presents a work with more narrative detail than historical analysis, but which provides useful context for the present research. In particular, the disadvantages of early photographic technology for purposes of newsgathering support the argument that police of the time did not need to fear their misconduct being recorded.

The value of Collins' (1990) work to the present research includes evidence that civilian cameras were in public space in numbers too large to control or ignore, that they were capable of capturing the actions and identity of police while remaining unobtrusive, and that an amateur photograph documenting a moment's indiscretion could end up in a national magazine.

Jacob Riis was a social reformer who had been a police beat reporter for the *New York Tribune* beginning in 1877 (Riis & Yochelson, 2001, pp. 6). As a reporter, he wrote 'the first draft of history'; as a reformer, and in his memoirs, he set his hand to later drafts. Of particular relevance to the present research, he visually documented the horrors of urban poverty with a basic dry plate camera, converted a hundred of his images into lantern slides, and gave his first two-hour illustrated lecture, "The Other Half, How It Lives and Dies in New York" in 1888 (Hales, 1984, p. 176). Scribner's published a book-length collection of his images and lectures in 1890 (Riis & Yochelson, 2001, pp. 7-9). Riis was often accompanied by police as he took photographs, and police are evident in some of his photos. Theodore Roosevelt met Riis in 1894, and was influenced at least in part by his photographs to put teeth into police reforms as President of the Police Board (Riis, 1901). Riis' first-hand accounts of police response to photography, of police misconduct and corruption, and of the efforts of police reformers such as Roosevelt, make these works valuable in developing the historical context of the present research.

Hales (1984) makes a number of observations on the social effects of the combination of the Kodak camera, the rise of Progressivism, and the development of cheap halftone technology for newspapers. This is relevant to the present research for the documentation of historical precedent to user-generated online video as a tool for social change.

In *Watching YouTube*, Strangelove (2010) discusses YouTube as a new channel through which anyone can transmit their messages. He makes four major distinctions between YouTube and previous media. First, YouTube is a participatory medium which enables the representational power of creators who do not have access to traditional broadcast or print media. Second, YouTube is an interactive medium which enables direct commentary and response from the audience with a reach and exposure equal to that of the original message, in contrast to the absent or extremely limited, generally asynchronous, and significantly reduced reach of most mass media audience responses. Third, YouTube has very little mediation, unlike the ownership and editorial mediations of traditional media. Finally, YouTube is a relatively transparent medium, in which the creators of a message – particularly one of propaganda – who attempt to conceal the message's origins may be exposed and counter-programmed by smaller organizations or even a sufficiently motivated individual. Astroturf – fake grassroots propaganda – is particularly vulnerable to exposure and ridicule through YouTube mash-ups (Strangelove, 2010). As will be detailed later, the author's discussion is highly relevant to the present research.

RELEVANT LEGAL ISSUES

Even as technology has made cameras more numerous and ubiquitous, cultural factors in the United States have increased sensitivity to cameras. Even those who choose not to use cameras are likely to be visible to the cameras of others. Many people are unclear on what is or is not legal: a photojournalist on assignment, a homeowner placing a security camera, a private security guard attempting to enforce a company policy, a cell phone user capturing a public incident, a parent whose child is being imaged, a police officer responding to a complaint or making an arrest. To avoid needless conflict, it is important for people to understand both the laws affecting their use of cameras and the laws affording protection from the cameras of others. As of this writing, a synthesis of this information does not appear to have been gathered and examined in one place.

The scope of the following review is narrow in discipline, jurisdiction, technology and depth. The discipline of this review is law; it does not include technical or social history except as it directly affects the law, and only considers public space in areas subject to U.S. law. In accordance with the history of imaging technology, there is little citation of law prior to 1839. This review considers any imaging technology, particularly including still photography, cinematography, and videography, but does not address audio recording except as it is an inextricable part of audiovisual recording. This is intended to be a brief overview, and therefore does not go into exhaustive detail; it is intended to function as a road map to the history of the current laws, regulations, policies and common practices affecting the use of imaging technology in American public spaces.

A single chronology of case citations is not sufficient to address the complexity of the issues affecting the photography of public spaces. The law is a balancing act, influenced by changes in technology, public mores and other factors; a weight on one side of an issue may not be balanced for years or decades. Accordingly, this report presents each issue in its own chronology, and the issues are presented in relative order of importance in the hope that the reader can more easily make sense of the whole.

Photography as Protected Speech

The primary law regarding photography of public spaces is the First Amendment to the Constitution of the United States, specifically the prohibition against laws "abridging the freedom of speech, or of the press" (USCS Const. Amend. 1, 1789). Between 1789 and 1839 there was little legal interest in photography aside from patent suits among the various inventors, as photography was a collection of more-or-less successful experiments. The 1839 introduction to the United States of the daguerreotype, the first truly practical photographic process, did not materially change the situation. The process of taking a photograph still required enough time, effort and equipment that it could not be done quickly or surreptitiously, so nearly all photographs were formally posed and made with the consent of those pictured. Thus, there were no legal challenges to the taking of photographs during this time. The new technology was in fact praised for its utility, and one early source predicted its use for criminal justice: "What will become of the poor thieves, when they shall see handed in as evidence against them their own portraits, taken by the room in which they stole, and in the very act of stealing!" (Willis, Porter & Talbot, 1839, p. 71)

The first cited ruling as to the right to take photographs held that private property owners had the right to exclude or prohibit photography on their premises, but that they had no right to stop photography from outside the property (*Sports and General Press Agency v. "Our Dogs" Publishing Co.*, 1916). In the next significant ruling, *Humiston v. Universal Film Manufacturing Co.* (1919), the court held that moving picture newsreels are the same as newspapers for presentation of current events, and that newsreels are not purely commercial as is advertising. Because a newsreel is "a truthful picture taken of a current event at the time that it happened" (p. 2) there were no damages awarded for the plaintiff, an attorney who was pictured in public in the course of her work with the police on a case of current interest.

In 1968 during the Democratic National Convention, Chicago police allegedly refused to identify themselves, removed their badges, prevented photographers from taking news photos, and threatened the photographers. The City of Chicago, the superintendent of police, and unnamed police officers were sued in a class action by the president of the local chapter of the American Society of Magazine Photographers and other members of the press "requesting a permanent injunction to prevent city officials and police from interfering with photographers' right to report on and to photograph news events" (*Schnell v. City of Chicago*, 1969, pp. 1-2). The district court denied jurisdiction under 42 U.S.C.S. § 1983 and, on its own motion, dismissed the suit. The appeals court ruled that photographers are legitimate members of the press, and that photography is a constitutionally protected activity; it also reversed the district court ruling, and remanded for trial:

> From the literal wording of 42 U.S.C.S. § 1983, injunctive relief is a proper remedy if the alleged unconstitutional deprivation of rights is established. Under § 1983, equitable relief is appropriate in a situation where governmental officials have notice of the unconstitutional conduct of their subordinates and fail to prevent a recurrence of such misconduct. (*Schnell v. City of Chicago*, 1969, p. 3)

The courts have also taken notice of what photographers may not do. In particular, the courts have held that First Amendment protections do not immunize photographers against the consequences of criminal action. In a case where a journalist used a hidden camera without consent and under false pretenses to photograph the plaintiff in his own home, the court observed:

> The First Amendment has never been construed to accord newsmen immunity from torts or crimes committed during the course of newsgathering. The First Amendment is not a license to trespass, to steal, or to intrude by electronic means

> into the precincts of another's home or office. (*Dietemann v. Time*, 1971, p. 13)

The following year, the Supreme Court acknowledged the right of the press to gather information, since "without some protection for seeking out the news, freedom of the press could be eviscerated" (*Branzburg v. Hayes,* 1972, p. 681). This ruled that newsgathering, including photography, is on equal footing with news publication as a First Amendment right.

Additionally, the courts have held that the act of taking a photograph is an integral part of the process of free speech. In a case where a police sergeant said, "No pictures!" and seized a TV station's camera at the scene of an arrest, the court ruled that:

> ...films are subject to the protection of the First Amendment. ...it is clear to this court that the seizure and holding of the camera and undeveloped film was an unlawful "prior restraint" whether or not the film was ever reviewed. (*Channel 10, Inc. v. Gunnarson*, 1972, p. 7)

First Amendment protections are particularly sensitive to efforts by the state to restrict the content of protected speech, e.g., the subject matter of photographs. The courts have held that government action that acts to chill or repress free speech is undesirable and must withstand strict scrutiny. "For the state to enforce a content-based exclusion it must show that its regulation is necessary to serve a compelling state interest and that it is narrowly drawn to achieve that end" (*Perry v. Perry*, 1983, p. 45).

The United States does not license journalists or photographers, and courts have held that any person has the same free speech rights in making video recordings as a member of the press. In one case, police had seized and refused to return a videographer's tape of a fatal gang fight, which he intended to sell to a television station: "It is not just news organizations, such as WHO-TV, who have First Amendment rights to make and display videotapes of events -- all of us, including Lambert, have that right" (*Lambert v. Polk County, Iowa*, S.D. Iowa 1989, p. 12).

When courts have considered restrictions on First Amendment protected activities, legitimate state interests have been ruled to include public safety (*Dayton Newspapers v. Starick*, 1965, pp. 3-4), ensuring the flow of traffic (*Perry v. LAPD*, 1997, p. 4), maintaining the orderly movement of pedestrians (*Heffron v. Krishna*, 1981, p. 650), and preventing interference with an investigation of physical evidence (*Gazette v. Cox*, 1967). However, simply asserting 'public safety' is not sufficient for limiting media access; there must be substantiating evidence (*Gannett v. Pennsauken*, 1989, pp. 536-537).

Police cannot with impunity put a hand over a lens or shove the camera into the journalist's face. An amateur journalist was arrested in the course of videotaping a public protest, and thereafter brought suit against the city and the officers under 42 U.S.C. § 1983 for interfering with "his First Amendment right to film matters of public interest" (*Fordyce v. City of Seattle*, 1995, p. 5). The district court, among other rulings, concluded that there was no evidence of the officer's alleged assault and battery of the journalist. The appeals court disagreed: "Fordyce's allegation is nonetheless corroborated by his videotape, which is in the record and which we have reviewed" (p. 6). The appeals court reversed and remanded.

Police in Cummings, Georgia allegedly prevented James Smith from videotaping their actions. The district court denied that Smith had a right to do so. The appeals court confirmed that the appellant private citizen had "a First Amendment right, subject to reasonable time, manner and place restrictions, to photograph or videotape police conduct. The First Amendment protects the right to gather information about what public officials do on public property, and specifically, a right to record matters of public interest" (*Smith v. City of Cummings*, 2000, pp. 2-3).

Not all photography is equal under the First Amendment. The strongest protection applies to imaging of public interest, and to imaging intended for publication. "...in order to be protected under the First Amendment, images must communicate some idea" (*Porat v. Lincoln Towers,* 2005, p. 13). In this case, the plaintiff identified himself as a 'photo hobbyist' and stated that his interest in photographing high-rise buildings from private property (where he was ticketed for trespass) was only for 'aesthetic and recreational' purposes. These purposes are not enough to satisfy the 42 U.S.C.S. § 1983 First Amendment retaliation standard, and the plaintiff's claims were dismissed.

In one of the cases examined in the present research, the U.S. Court of Appeals for the First Circuit in Boston ruled that "a citizen's right to film government officials, including law enforcement officers, in the discharge of their duties in a public space is a basic, vital and well-established liberty safeguarded by the First Amendment" (*Glik v. Cunniffe et al.*, 2011e, p. 17; Waterman, 2011). The court also concluded that "Glik was exercising clearly established First Amendment rights in filming the officers in a public space" (*Glik v. Cunniffe et al.*, 2011e, p. 2).

Public Space, Place, Forum or Venue

The most significant limitation on First Amendment protections for photography is the place from which the photograph is taken. Public space has some of the strongest protections; photography on someone else's private property can be much more restricted.

One of the earliest rulings on this difference was handed down in 1933. The plaintiff sued a newspaper for printing a photograph of her, and the newspaper's defense was partially based on the public place in which the photo was taken. The court held:

> The plaintiff's allegations show that the picture of which she complains was not taken surreptitiously or without her knowledge and consent. On the contrary she voluntarily posed for it as one of the party of five. The picture was taken at an airport which is presumably a public place. One who under the conditions disclosed in these counts poses for a photograph has no right to prevent its publication. (*Thayer v Worcester Post Co.*, 1933, pp. 163-164)

The tradition of the public forum as the home of public discussion and the marketplace of ideas is central to democracy, and therefore has some of the strongest protections for freedom of expression. The term "public forum" as used by the courts refers to a public place that has been used historically as a venue for free expression. In *Hague v. CIO* the court held that "…the public places of a city must be open for the use of the people in order that they may exercise their rights of free speech and assembly" (*Hague v. CIO*, 1939, p. 32).

In a ruling similar to that twenty years earlier, the court ruled in *Gill v. Hearst Publishing Co.* that the public forum in which a photograph was taken invalidated the basis of their claim:

> Here plaintiffs, photographed at their concession…in the Farmers' Market, had voluntarily exposed themselves to public gaze in a pose open to the view of any persons who might then be at or near their place of business. By their own voluntary action …plaintiffs' right to privacy as to this photographed incident ceased and it in effect became a part of the public domain. (*Gill v. Hearst Publishing Co.*, 1953, p. 230-231)

In a third case, a child's body was found in public view, although on private land, and the published photograph of the body was the stimulus for a suit by the family. The two facts of public view and the story being newsworthy were sufficient for the court. (*Bremmer v. Journal-Tribune Publishing Co.*, 1956)

Courts have ruled that newsgathering, including photography, is protected speech in public forums. "…employees of the news media have a right to be in public places and on public property to gather information, photographically or otherwise" (*Channel 10, Inc. v. Gunnarson*, 1972, p. 11).

However, not every public space is necessarily a public forum. For example, a sidewalk, even if owned by the government, is not a public forum if it is on a military base. Even if the general public is occasionally invited or permitted to use such a space, that does not make the space a public forum. "The public interest in insuring the political neutrality of the military justifies the limited infringement on First Amendment rights imposed by Fort Dix authorities" (*Greer v. Spock*, 1976).

The government may reasonably limit photography and other First Amendment activities in a public forum for valid reasons. For example, Minnesota Fair rules confined expressive activities to specific, fixed areas (booths) at a large state fair to avoid congestion and maintain the orderly movement of fairgoers. The court held that "a State's interest in protecting the 'safety and convenience' of persons using a public forum is a valid governmental objective" (*Heffron v. Int'l Soc. for Krishna Consciousness, Inc.*, 1981).

Public space has been consistently held by the courts to include streets, sidewalks, parks, and other public property. It is difficult for government agents to legally justify restricting photography in or from these public spaces. The Supreme Court considered limiting First Amendment-protected expression on its own sidewalks in *United States v. Grace*, (1983). "The sidewalks comprising the outer boundaries . . . are indistinguishable from any other sidewalks in Washington, D.C., and we can discern no reason why they should be treated any differently" (p. 177). Thus, the court held consistently with precedent that:

> In such places, the government's ability to permissibly restrict expressive conduct is extremely limited; the government may enforce reasonable time, place, and manner regulations as long as the restrictions 'are content-neutral, are narrowly tailored to serve a significant government interest, and leave open ample alternative channels of communication.' (p. 177)

However, courts have held that "public property which is not by tradition or designation a forum for public communication is governed by different standards" (*Perry Educ. Ass'n v. Perry Local Educators' Ass'n*, 1983, p. 45). A school's internal mail system, although available to some members of the public for specific purposes, was not held to be a public forum for First Amendment purposes.

Courts have distinguished traditional public forums as having greater First Amendment protection than other public spaces including prisons, military installations, airports, border crossings and courthouses (*Wilkinson v. Frost*, 1987).

The state cannot interfere with protected expression in public forums simply by invoking public safety. "Although safety is undoubtedly a significant government interest, broad assertions of a safety interest, without evidence to substantiate them, cannot survive when the First Amendment is implicated" (*Gannett v. Pennsauken*, 1989, p. 19).

Court rulings regarding the definition of traditional public forums have not always been clear. In *United States v. Kokinda*, (1990), the Supreme Court was more divided than usual. A sidewalk outside a post office but entirely on postal service property and used for no other purpose was interpreted by various Justices to be a traditional public forum, a nonpublic forum, "more than a nonpublic forum" (p. 737) and a "limited-purpose public forum" (p. 752). Significant factors were that the plaintiffs had been soliciting funds, had set up a table blocking most of the narrow sidewalk, and that over forty postal customers had complained. The postmaster asked plaintiffs to leave; they refused; they were arrested by postal inspectors and charged with violating a statute prohibiting solicitation of funds on postal premises. In a five-to-four decision, the Court held that the plaintiff's First Amendment rights had not been violated – but the Court never did agree on the legal definition of that sidewalk.

If the state seeks to limit protected speech activities such as photography, those limits cannot be overly broad and must be narrowly tailored to a significant government interest (*Perry v. Los Angeles Police Dep't*, 1997; *Warren v. Fairfax County*, 1999; *United for Peace and Justice v. City of New York*, 2003; *Black Tea Soc. v. City of Boston*, 2004). Even a limited public forum is protected, particularly against content-based restrictions, and requires narrowly tailored limits that serve a compelling state interest (*Burnham v. Ianni*, 1997).

It is particularly difficult for government agents to legally restrict photography from one's own private property or from private property where the photographer has express permission from the owner, even in cases of wrongful conduct by the photographer. In *Balboa v. Lemen* (2004) the trial court found that the defendant had regularly videotaped and photographed patrons of a neighboring restaurant (including following them to their cars in the business's parking lot and flash photography at night through the business's windows), thereby creating a nuisance. The trial court, upheld by the appeals court, granted a permanent injunction against Lemen:

> To the extent [this injunction] affects Lemen's free speech rights, it is reasonable in scope, clear, and 'sweeps no more broadly than necessary' to abate the nuisance. [It] permits Lemen to take photographs from more than 25 feet from the Village Inn premises, from her own property, or to document disturbances or damage to her property. (*Balboa v. Lemen,* 2004, pp. 604-605)

Clearly, the location from which a photographer is working can make a significant difference in how much (if any) First Amendment protection applies to that activity. The photographer's actions or position within the location can also affect the protection afforded.

Privacy

A second area of limitation to First Amendment protections for photography is privacy. Privacy as a legal concept in America is not originally based on a particular statute, and the word does not appear in the U. S. Constitution or any of its amendments. Most of the existing case law is based on a series of opinions in an article written by Samuel Warren and Louis Brandeis and published in *The Harvard Law Review* in 1890 (Warren & Brandeis, 1890). The impetus to this article was the use by the 'yellow press' of the newly developed Kodak camera. Previously, the process of taking a photograph was time-consuming and laborious enough that it was nearly impossible to take a person's picture without permission. In these circumstances, the law of contract or of trust was enough to safeguard a reasonable person against misuse of their likeness. However, the Kodak 'instant' camera made surreptitious or candid photography possible. Unscrupulous journalists used the new technology to take pictures of members of high society. The key incident appears to be coverage in Boston's *Saturday Evening Gazette* of the wedding celebrations of a daughter of Mr. and Mrs. Warren in 1890. In that city at that time and in that social circle, such publication was enough to annoy Mr. Warren (Prosser, 1960, p. 383). He and his recent law partner Brandeis wrote, "Instantaneous photographs and newspaper enterprise have invaded the sacred precincts of private and domestic life; and numerous mechanical devices threaten to make good the prediction that 'what is whispered in the closet shall be proclaimed from the house-tops" (Warren & Brandeis, 1890, p. 195).

Warren and Brandeis went on to develop the legal theory that "...the doctrines of contract and of trust are inadequate to support the required protection, and the law of tort must be resorted to" (Prosser, 1960, p. 383). A *tort* is "a civil wrong against another that results in injury" (RCFP, 2007a). Note that a tort cannot exist until an act has been committed; the tort itself is only useful in a civil suit for damages after the fact. As proposed by Warren & Brandeis and developed by subsequent scholars and rulings, there are four privacy torts providing for civil action for redress: publication of private facts, intrusion, false light, and commercialization (Middleton & Lee, ch. 5). In practice, this means that privacy concerns are not grounds for preventing photography, especially in public spaces. Absent federal statues, privacy statutes have developed at the state level.

The first court ruling recognizing a privacy tort came just two years later. An adjoining property owner sued the operators of an elevated railway because the location of a station platform and stairs exposed her tenant's third-floor windows, previously private, to the view of the passengers and crew of the railroad. The court recognized that privacy, a property, had been taken, and that the loss was evident in lost rent and lowered property value. The plaintiff was compensated under eminent domain (*Moore v. New York Elevated R. Co.*, 1892).

Not all courts fell into line with Warren & Brandeis immediately. Ten years after *Moore*, a company used unauthorized and uncompensated photographs of a person in advertising materials, and the court refused to recognize any privacy tort (*Roberson v. Rochester Folding Box Co.*, 1902). This ruling outraged enough people that the New York state legislature enacted a privacy statute (NY Sess. Laws ch. 132, §§ 1-2, 1903) making it both a misdemeanor and a tort to use a person's photograph for trade without consent (Prosser, 1960, p. 385). The first appellate recognition of the privacy doctrine of Warren & Brandeis came two years later when the Georgia supreme court ruled that publication of a person's photograph without consent, purely for commercial gain, is an invasion of privacy (*Pavesich v. New England Life Ins. Co.*, 1905).

In a 1908 ruling along similar lines to *Moore,* the court found that because prisoners in the newly constructed county jail were able to look into the appellant's home windows, the "appellant's right of privacy has been invaded" (*Pritchett v Board of Commissioners of Knox County*, 42 Ind. App. 3 1908).

The ruling in *Humiston v. Universal Film Manufacturing Co.* (1919) is particularly significant for photography because the court subordinated the right of privacy to the doctrine of the freedom of the press. The court based the ruling on the public space location of the original photography, and on the court's interpretation of the moving picture newsreel being the same as newspapers rather than being purely commercial as is advertising (Rice, 1920, pp. 286-287).

The ruling in *Barber v. Time, Inc.*, (1942) established that even the apparent newsworthiness of a photograph does not override all considerations of privacy. In this case, the plaintiff was photographed without her consent in a hospital room while she was being treated. The court ruled that she did have an expectation of privacy, that the magazine had violated that privacy by publishing the picture, and that the fact that the story was newsworthy did not immunize the magazine against a privacy tort claim.

The use and re-use of news photographs has led to some complex rulings. A photograph of a child involved in a street accident, where the driver was at fault, was published in the local newspaper the following day. The court recognized the public space reduction of an expectation of privacy and the newsworthiness of the subject matter; a privacy tort claim on the initial publication was not sought by the plaintiff. However, the photograph was acquired and re-published in the *Saturday Evening Post* twenty months later in an article on pedestrian carelessness entitled, 'They Ask To Be Killed.' The court ruled that the second publication was:

> ...an actionable invasion of plaintiff's right of privacy. Granted that she was 'newsworthy' with regard to that particular accident for an indefinite time afterward. This use of her picture had nothing at all to do with her accident. ...the little girl, herself, was at the time of her accident not careless and the motorist was...this particular plaintiff...now becomes a pictorial, frightful example of pedestrian carelessness. This, we think, exceeds the bounds of privilege. (*Leverton v. Curtis Pub. Co.*, December 1951, p. 11)

The appeals court affirmed the judgment for the plaintiff of $5000. However, the district court had observed that "The right [of privacy] is, of course, variable and in some cases it may dwindle almost to the vanishing point" (*Leverton v. Curtis Pub. Co.*, March 1951, p. 2).

State privacy law continued to develop, and some states passed what came to be known as "Peeping Tom" statutes. These had implications for both privacy tort claims and criminal law, as illustrated in *Souder v Pendleton Detectives*, (1956). The trial court in Louisiana maintained that there was no cause of action, but the appeals court ruled that:

> ...it appears that the detectives might have been guilty of a crime under our 'Peeping Tom' statute. LSA-R.S. 14:284 defines a 'Peeping Tom' as 'one who peeps through windows or doors, or other like places, situated on or about the premises of another for the purpose of spying upon or invading the privacy of persons spied upon without the consent of the persons spied upon. It is not a necessary element of this offense that the 'Peeping Tom' be upon the premises of the person being spied upon.' (*Souder v Pendleton Detectives*, 1956, p. 4)

The appeals court reversed and remanded. It is worth noting that a 'Peeping Tom' or anti-voyeurism statute may impose criminal penalties for photography from a public space or one's own private property. This is a complex area of state law, beyond the scope of this work, and warrants careful study of the particular jurisdiction before engaging in photography that may be interpreted to be criminal.

The body of privacy case law - over three hundred cases as of 1960 - and Warren & Brandeis' original doctrine were examined by William L. Prosser in his law review article "Privacy," much of which was incorporated later into *The Restatement (Second) of Torts*. These works are widely cited, and form a crucial part of photography law. A particularly important passage is:

> On the public street, or in any other public place, the plaintiff has no right to be alone, and it is no invasion of his privacy to do no more than follow him about. Neither is it such an invasion to take his photograph in such a place, since this amounts to nothing more than making a record, not differing essentially from a full written description, of a public sight which any one present would be free to see. (Prosser, 1960, pp. 391-392)

Prosser followed this immediately by the warning, "...when he is in the seclusion of his home, the making of a photograph without his consent is an invasion of a private right, of which he is entitled to complain" (Prosser, 1960, p. 392).

Prosser explicated in detail the four privacy torts: intrusion, public disclosure of private facts, false light, and misappropriation. These torts had been accepted by the courts in at least 28 states at the time, as well as in the *Restatement (Second) of Torts* (Prosser, 1960, pp. 391-392).

The right to take pictures in a public place was reaffirmed regarding both privacy and the idea of property rights: "anything visible in a public place may be recorded and given circulation by means of a photograph...since this amounts to nothing more than giving publicity to what is already public" (Prosser, 1960, pp. 394-395). Prosser discussed moving pictures as well as still photography, particularly as affected by the tort of misappropriation:

> It has been held that the mere incidental mention of the plaintiff's name in a book or a motion picture or even in a commentary upon news which is part of an advertisement, is not an invasion of his privacy; nor is the publication of a photograph or a newsreel in which he incidentally appears. (Prosser, 1960, pp. 405-406)

In 1963 an appellate court held that simply following a person and filming their actions in public did not constitute an invasion of privacy, even though the plaintiff was upset by the surveillance. A detective had been hired by the plaintiff's insurance company to ascertain the accuracy of the plaintiff's claims of disability. The court noted that all surveillances took place in public spaces, where only limited expectation of privacy applies. The court also ruled that:

> Although the so-called "public figure" limitation upon the right to privacy has generally been applied to such persons as actors, public officials, and other newsworthy persons, its rationale also applies to a person who makes a claim for personal injuries. (*Forster v. Manchester*, 410 Pa. 192, 150 (Pa. 1963))

The court ruled that the detectives were simply doing their jobs, which had social value, and they did not intend to cause emotional distress. The plaintiff was therefore denied redress.

Not everything that happens in a public place is fair game for a photographer. A mother with her two young sons leaving a 'fun house' at a county fair did not know about the air jet positioned to blow skirts upward, and was caught unawares. A local newspaper photographer, waiting for just such an opportunity, took her picture at that moment, and the paper published it on its front page without her knowledge or consent. Because of this publication, the plaintiff "became embarrassed, self-conscious, upset and was known to cry on occasions" (*Daily Times Democrat v. Graham*, 1964, p. 5). The court observed, "We can see nothing of legitimate news value in the photograph. Certainly it discloses nothing as to which the public is entitled to be informed" (*Daily Times Democrat v. Graham*, 1964, p. 8). The court also noted that the plaintiff was not a public person. To the defense of the photograph being taken in a public place, the court replied: "To hold that one who is involuntarily and instantaneously enmeshed in an embarrassing pose forfeits her right of privacy merely because she happened at the moment to be part of a public scene would be illogical, wrong, and unjust" (*Daily Times Democrat v. Graham*, 1964, p. 10). The appeals court affirmed the jury award to plaintiff of $4,166 in damages.

Fourth Amendment protections against unreasonable searches have also been invoked to protect privacy against government intrusion, notably in *Katz v. United States*. The court held that the Fourth Amendment "protects people, not places," and that

What a person knowingly exposes to the public, even in his own home or office, is not a subject of Fourth Amendment protection… [but] what he seeks to preserve as private, even in an area accessible to the public, may be constitutionally protected. (*Katz v. United States*, 389 U.S. 347 (1967))

Following the Katz decision, Congress passed Title III of the Omnibus Crime Control and Safe Streets Act of 1968 with the intent of protecting individuals from uncontrolled electronic surveillance by either law enforcement or civilians. According to the new law,

> No violation of the anti-wiretapping statute exists if one or more parties to the recording consents, one party lacks a reasonable expectation of privacy in the conversation, or if a warrant was procured in good faith by a law enforcement official. (Skehill, 2009, p. 989)

However, court rulings on matters of photographic technology are not always clear-cut and consistent. One court held that police use of a telephoto lens to gather evidence was not a violation of privacy, partly because the defendant was in another person's driveway and therefore had no expectation of privacy (*Michigan v Ward* 107 Mich. App. 38 (1981)). Similarly, courts have held that there is no reasonable expectation of privacy if activities are visible from low-flying aircraft, including looking through greenhouse roof vents (*Florida v. Riley* 488 U.S. 445 (1989); *Dow Chem. Co. v. United States*, 476 U.S. 227, 238 (1986); *California v. Ciraolo*, 476 U.S. 207 (1986)). "Where a telephoto lens has been used to view activities outside a residence, other courts have found no error" (*United States v Gibson*, U.S. App DC; 636 F2d 761 (1980)), (*United States v Allen*, 633 F2d 1282 (CA 9, 1980), *United States v Minton*, 488 F2d 37, 38 (CA 4, 1973), *United States v Grimes*, 426 F2d 706 (CA 5, 1970)). However, rulings have increasingly found that high-powered lenses viewing private spaces through windows have violated privacy. Other courts have held that thermal imaging that shows personal activity inside a home is an unreasonable search and a privacy violation, but that thermal imaging that only shows an external temperature difference is not a search (*Kyllo v. United States*, 533 US 27 (2001)). If a person wishes to legally secure their strongest guarantees of privacy, they must make an effort to enclose private activities within the curtilage of the home, including protecting the area from view of passersby (1987 *United States v. Dunn*, 480 U.S. 294 (1987)).

On the other side of the privacy issue, state-by-state anti-voyeurism laws have been revised over time, and presently 'upskirt' laws are being passed in some states. In Maine,

> A hidden cameras law makes it a "Class D" crime to use a camera to view or record a person in a private place, "including, but not limited to, changing or dressing rooms, bathrooms and similar places," or in a public place if one views any portion of another person's body "when that portion of the body is in fact concealed from public view under clothing," and a reasonable person would expect it to be safe from surveillance. Me. Rev. Stat. Ann. tit. 17-A, §511. (Reporters Committee for Freedom of the Press, 2008, para. 4)

Imaging that includes sound may come under the stronger privacy restrictions of wiretap law, including the requirement for consent. "Journalists should be aware, however, that the audio portion of a videotape will be treated under the regular wiretapping laws in any state" (Reporters Committee for Freedom of the Press, 2008, para. 9). Another issue is that of concealment; if the camera is clearly visible, a statute specifically against concealed devices does not apply. "[T]he district court dismissed the criminal charges against Glik because, unlike Hyde, Glik had been holding his cell phone in plain view" (Skehill, 2009, pp. 983-984; *Massachusetts v. Glik*, No. 0701 CR 6687, slip op. at 3-4 (Boston Mun. Ct. Jan. 31)).

The courts have consistently held that public officials in the performance of their duties in a public place have no expectation of privacy. "The court held that the arrest was not entitled to have been private and the officers could not reasonably have considered their words private. Because the exchange was not private, its recording could not have violated § 9.73.030" (*Washington v. Flora*, 845 P.l2d 1355, 1358 (Wash. 1992)). Similarly,

The officers could not maintain a civil action under the New Jersey Wiretapping and Electronic Surveillance Act, N.J. Stat. Ann. §§ 2A:156A-1 to -34, because they had no reasonable expectation of privacy in the car they were searching as they were recorded. Finally, the police could not recover damages for fraudulent news gathering, because the newsmagazine was not spying on private matters, but observing public officials in the performance of their duties. (*Hornberger v. Am. Broad. Cos.*, 799 A.2d 566, 626-27 (N.J. 2002))

In some jurisdictions, a series of conflicts over interpretation of privacy statutes and civilian recording of police has led to clarification of policies in line with the law. For example,

> Police in Spring City and East Vincent Township [Pennsylvania] agreed to adopt a written policy confirming the legality of videotaping police while on duty. The policy was hammered out as part of a settlement between authorities and ACLU attorneys representing a Spring City man who had been arrested several times last year for following police and taping them. (Rowinski, 2010)

Private security or law enforcement officers have been reported to cite the PATRIOT Act when challenging photographers. That act does not restrict imaging of public spaces beyond existing statutes. "No specific post-September 11 federal law grants the government any additional rights to restrict visual newsgathering, photojournalism or photography generally" (NPPA, 2005). Furthermore, the courts have held that simply taking a picture is not grounds for an action; the image must be used in a way that injures the person pictured, according to one of the four privacy torts. One of the positive defenses to a private-facts tort claim is that the subject of the recording is "of legitimate concern to the public" (Restatement (Second) of Torts, § 652D, 1977; *Shulman v. Group W. Prods.*, 955 P.2d 469 (Cal. 1998)).

Since the U.S. Court of Appeals for the First Circuit in Boston ruled on *Glik v. Cunniffe et al.*, a number of law reviews have cited that case in examining police privacy and citizen recording issues in detail (Brncik, 2011; Potere, 2012; Van Tassell, 2013; Harvard Law Review, 2013b). The case has also figured prominently in *Sharp v. Baltimore City Police et al.*, particularly in the Statement of Interest filed by the US DOJ (*Sharp v. Baltimore City Police Department, et al.,* 2012).

Fourth Amendment

Protections against search and seizure apply to photographer's equipment, film and recording media. Furthermore, copyright law states that once a work is created in a fixed form, it is copyrighted; when you take a picture, you create an original work that has value and that belongs to you. A photograph or video recording has been held to be intellectual property (*Lambert v. Polk County, Iowa*, S.D. Iowa 1989), with Fourth Amendment protections. Police wishing to examine that image, destroy it, or to take it from you, must follow due process:

> The right of the people to be secure in their persons, houses, papers, and effects, against unreasonable searches and seizures, shall not be violated, and no Warrants shall issue, but upon probable cause, supported by Oath or affirmation, and particularly describing the place to be searched, and the persons or things to be seized. (USCS Const. Amend. 4)

In one of the cases examined in the present research, the U.S. Court of Appeals for the First Circuit in Boston concluded that the citizen photographer "was exercising clearly established First Amendment rights in filming the officers in a public space, and that his clearly-established Fourth Amendment rights were violated by his arrest without probable cause" (*Glik v. Cunniffe et al.*, 2011e, p. 2).

ADOPTION OF PHOTOGRAPHIC TECHNOLOGY, 1839-1979

In order to consider the interaction of police, civilians, and cameras in American public space, it is necessary to understand where the cameras came from and the camera's place in American society. The following section surveys the history of imaging technology, with the goal of explaining where all these cameras came from, what they have cost, what American social and individual attitudes have been towards them, and their technological capabilities and limitations up to the present day, much as Taft (1938) "endeavored to trace, however imperfectly, the effects of photography upon the social history of America, and in turn the effect of social life upon the progress of photography" (p. viii). This is a historiography for most of the period examined, with the most recent data drawn from primary sources.

The scope of this section is limited to chemical imaging cameras - that is, the technology of paper and film. The period examined is from 1839, the first published American editorial on photographic technology (Willis, Porter & Talbot, 1839), to the cessation of Kodak's and Bell & Howell's American manufacture of amateur movie cameras in 1979, which signaled the end of film cameras' public heyday and the transition to electronic cameras (*The New York Times*, 1981).

Chemical photographic technology and the societal changes interacting with it are firmly embedded in the American consciousness and language. The end result of the maturity of chemical photographic technology was a wide selection of affordable, capable, full-featured cameras, and the knowledge that even a child could use them to produce consistently good pictures. Nearly every American household in the early 1870s owned at least one photograph, (Hales, 1984, p. 5) and Oliver Wendell Holmes described the contents of the family photographic album as "the social currency, the sentimental "green-backs" of civilization" (Taft, 1938, pp. 143). This widespread ownership of photographs began the generational familiarization that prepared the family to use the camera as well, without the intermediary of the professional photographer. A century later, new camera models could be expected to sell tens of millions of units, and it was a rare household that did not own at least one camera (Collins, 1990, p. 310).

That photography is inextricably part of American history is evident from its traces on the American vernacular. The phrase "camera shy" first entered the American lexicon in print, apparently coined at the time by Frederic Hart Wilson in an article entitled "Some Comments on the German Group" in 1890: "If they are gallery-scared, 'camera-shy' to borrow a metaphor from the sportsman, he must try to get them interested, and to make them – prime requisite – forget themselves" (Wilson, 1890, p. 346). Like the term snapshot (Collins, 1990, p. 72), it was borrowed from the language of firearms. The term "shutterbug" did not appear in print until 1940, but as it then appeared in a book title, newspaper advertisements, and a language journal contemporaneously it had probably been in common use for some years before then (Johnston, 1940, 357). Such vernacular terms speak directly to a number of the case studies examined in this monograph.

The history of chemical photographic technology in America can be usefully divided into three phases: pioneering, popularization, and maturity. Pioneering as a historical phase can be characterized as one of experimentation and invention, the trial of competing approaches to each problem. This phase was dangerous, both physically and economically, for those brave enough to participate. The technologies were hazardous, and most of the hazards were not recognized as such until it was too late (Carlebach, 1992, pp. 10-11, 30-31; Taft, 1938, p. 59). For the investor or inventor, the return on investment of money, time and labor was almost certain to be nil; there were more losers than winners in the game. Patents and licenses were invitations to sue or objectionable as to cost, and sales were uncertain at best in a yet tiny and fast-changing market (Taft, 1938, pp. 107-111, 375; Jenkins, 1975, pp. 160-161; Carlebach, 1992, pp. 7-9).

For the consumer, purchased goods were reliant on individual craftsmanship rather than mass production (Hales, 1984, p. 11), and as likely as not to be flawed, of poor design or uncertain quality. The individual's time might return pleasing results or not, for no reliable reason, as there was no functional theory to guide development in either chemical processes or mechanical design (Taft, 1938, pp. 382-383). Most of the products offered for sale were challenging to use, heavy, awkward, and potentially lethal. The expertise demanded combined that of a practicing chemist, optician, and mechanic – not to mention stevedore – with the sensibilities of a visual artist. Necessarily, the pool of those both able and interested was very small (Collins, 1990, p. 46). Fortunately for the future of photography, members of that set proved to be extremely enthusiastic and persistent – and were recognized as such (not always complimentarily) by their friends, family and associates (Taft, 1938, p. 374).

Popularization began with a marked turning point in ease of use. With that achieved, it became possible to market products to a wider segment of the population, which in turn made feasible both larger-scale scientific research and the tooling for mass production (Taft, 1938, pp. 122, 145). Standardization of materials and processes, combined with mechanical designs suited for assembly-line production using interchangeable parts (Collins, 1990, p. 50), made for more consistent quality in the product (both camera and film) and in the results for the photographer.

With these advances in production capacity, manufacturers could choose to exploit markets in several ways; they could continue to price goods for the upper-middle to upper classes (West, 2000, p. 75), they could reduce prices to exploit the larger market, or they could work out some combination of the two. This led to a market shakeout and differentiation, as the low-volume craftsmen either continued to serve a small clientele or went out of business, and the small number of photographic industrialists either expanded on their successes or went down in defeat (Taft, 1938, pp. 379-383). The most successful competitors are still American household names a century later, particularly Eastman Kodak and Bausch & Lomb.

Success in mass markets demanded lower production costs for materials, processes and durable products. Mass advertising became at least as important as word of mouth and reputation had been in the pioneering phase (Collins, 1990, pp. 57, 73). Products evolved rapidly under the stimulus of wider customer feedback, with economically safer opportunities for return on investment (Taft, 1938, p. 385-386; Collins, 1990, pp. 49-53). Physical safety was an unspoken concern; dedicated amateurs and professional photographers might tolerate corrosives, poisons and fire hazards, but the general public might not, so it was judged best not to mention those dangers while working behind the scenes to mitigate them (Katelle, 2000, pp. 52-61).

The most technically challenging tasks of the photographic process were taken out of the hands of the photographer and retained by the industrial or service sector specialists (Taft, 1938, p. 386). Ease of use became the watchword (Collins, 1990, p. 60), and led to exponential growth in sales, market size, and societal permeation. According to West (2000), "Kodak has done more than any other single enterprise or individual to determine the uses and expectations for snapshot photography, thereby also reshaping perceptions of such abstract concepts as memory and evidence" (p. xii).

Maturity in American chemical photographic technology was recognizable by indicators of success, particularly those of market saturation. The typical American home had at least one film camera. Every extended family had at least one member who wore the domestic label of "shutterbug," and perhaps more than one was recognized as "camera-shy." Advertising for cameras, supplies and developing services was a steady drone along with that for toasters and washing machines (Collins, 1990, pp. 82, 318-322; West, 2000). Entry for the new convert was attainable with a child's pocket money, and the paraphernalia of the advanced practitioner was available at the corner shop or the mail order catalog. Artists and professionals still argued arcana in public journals and private correspondence, and junior members of those circles could find guidance in local clubs or in classes offered by high schools, technical schools, colleges and universities. Self-study guides to photography were published in mass-market editions. Consumer expectations of point-and-shoot photographic technology were that you should always get at least a passable picture – unless you put your finger in front of the lens (Eastman Kodak, 1999; Collins, 1990, p. 309-310; Greenough, 2007, pp. 282-283). The basic technology did not change from year to year; any improvement seemed quite incremental, sometimes infinitesimally so.

In many ways, the evolution of motion pictures parallels that of still photography. Early motion picture cameras and their support equipment were heavy and bulky, and therefore not attractive to the general public. Hand-cranked motion picture cameras relied on the operator's skill to produce consistent speed of motion, and in some cameras proper exposure as well (Katelle, 2000, p. 52-3). The development of motors with governed speed output replaced that skill requirement, and was a significant step to wider acceptance. Film loading was a cumbersome process until film magazines or cassettes that could be loaded easily, and which protected the film from unwanted light exposure or manual misfeeding. Amateur cameras changed from 35mm to 16mm to 8mm film size, getting lighter, cheaper and easier to use. Film stocks became progressively more sensitive, enabling amateurs to shoot with available light, and to record sound on the film as well (Eastman Kodak, n.d.; Katelle, 2000, p. 273). Consumer expectations for home movie cameras became similar to those for point-and-shoot still cameras; even a child should be able to get passable footage. As Katelle (2000) commented, "Eastman Kodak's perfection of a system of film and equipment utilizing a narrow gauge of safety film made home movies immediately available to the merely well-to-do and, ultimately, within the reach of anyone who could afford to take snapshots" (p. vii).

Finally, a combination of new technologies and societal changes spelled the beginning of the end for chemical-based photography in America. Broadcast television and the incipient marketing of affordable videotape-based cameras cut the bottom out of the market for home movie equipment and film. This marked the beginning of the transition to broad public adoption of electronic imaging. By the time Kodak and Bell & Howell ceased American manufacture of home movie cameras in 1978, the Wolfman report for *Modern Photography* magazine estimated that seven million homes owned film projectors (*The New York Times*, 1981).

Social reactions to cameras appear to be remarkably consistent over time. The earliest cameras gave fair warning of the photographer's intent, and those who did not wish to be pictured were able to move out of the field of view or to remonstrate with the hapless shutterbug. The famous were known to complain of daguerreotypists who importuned en masse for commercial sittings, for there was a lucrative market in pictures of the prominent (Carlebach, 1992, p. 15). Apparently, the invention of the photograph was almost immediately followed by the innovation of the paparazzi. However, the average citizen did not have to be concerned about such attention until or unless they did something newsworthy. As Collins (1990) observed, "What is clear is that the world had become thoroughly seasoned to the culture of cameras" (p. 306).

Individual reactions to being photographed range from nonchalant to averse to incensed, in all phases of photography's development. Camera-shyness seems to have been present from the start, or perhaps earlier; snapshots of subjects with faces averted or covered are commonplace. Those who take most exception to being photographed seem to be those with a general concern for their privacy, rather than a specific dislike for the camera. One notable example of this class is Samuel D. Warren, a Boston businessman and Harvard Law School graduate. In 1890, angered by newspaper reports of his family's social entertainments, Warren coauthored with former classmate and law practice partner Louis D. Brandeis the *Harvard Law Review* article, "The Right to Privacy." This diatribe specifically takes aim against "instantaneous photographs" (Warren & Brandeis, 1890, p. 195), and successfully proposed a new set of four torts under which one could sue for violation of privacy. This is one of the most often cited law review articles in American jurisprudence.

ADOPTION OF ELECTRONIC IMAGING TECHNOLOGY, 1929-2010

By 2010, the American public had fully adopted electronic imaging technology with a range of expected features - namely, small enough to fit in a pocket or purse, weighing less than a wallet, operable by a child or a non-technical adult, capturing high-quality images – both still and video – easily and reliably in most environments, sharing that imagery wirelessly and through the Internet, self-contained yet compatible with external displays and printers, operating many hours continuously, recharging easily thereafter, and affordable for the vast majority.

American adoption of electronic imaging technology had followed a markedly different pattern from that of photographic technology, for a handful of reasons. First, photography from its beginnings produced a durable physical product that could be viewed by itself; video required complex and expensive viewing equipment, and did not have recording equipment for three decades after its original demonstration. Second, photography was simple enough that workable technique could be communicated to amateurs in a single newspaper article; building practical video cameras required teams of highly educated scientists and engineers. Third, photography produced excellent likenesses with fine detail from the beginning; video struggled for decades to render a recognizable likeness. Fourth, most of the successful photographic innovations were explicit goals achieved by purposeful effort; electronic imaging innovations tended to be adapted from byproducts of unrelated research. Fifth, commercial development of photography was based on sales of cameras, projectors, film, and developing services to the public for the creation of their own content; commercial development of video was based on a combination of sales of television receivers to a passive audience, and sales of advertising time to sponsors of network-produced content. There was simply no affordable, easy-to-use electronic parallel to the popular Kodak camera until the introduction of consumer camcorders and digital cameras in the last two decades of the twentieth century.

On the other hand, there were a few key similarities in the pattern of adoption that can usefully be categorized by generation of users. First, the generation-long familiarization following the expansion of regular television broadcasts after 1948, which conditioned Americans to 'look inside the frame' of the television receiver screen just as their great-grandparents had looked on early daguerreotypes. Second, the initially small but growing number of Americans who grew up with television, and who gained access to television cameras and made their own video recordings beginning in the late 1960s, 'filling the frame' just as George Eastman's contemporaries adopted and adapted photographic technology to their own purposes. Third, the rapid expansion into and permeation of American society by electronic cameras after visionary manufacturers (and their imitators) marketed the devices for the masses in the mid-1980s, paralleling Eastman's success with the Kodak and Brownie photographic cameras. Fourth, the generation of Americans born after 1985, who have never known a time without electronic cameras, who carry and use them without a second thought, and who regard 'filling the frame' with their own pictures and video as part of their technological birthright.

Magoun (2009) speaks to the present research more than any other scholar, interweaving social, political, economic, and technological history to clarify the many influences on the democratization of electronic imaging systems. Magoun (2009) also identifies corporate strategies that kept electronic cameras out of the hands of amateurs for decades. David Sarnoff's leadership of the Radio Corporation of America (RCA) continued the profitable monopolistic practice of buying up every competing patent it could not claim on its own. Sarnoff saw the commercial potential in centralized broadcast television based on his experience in radio (Magoun, 2009, pp. 51-53), and "RCA established a virtual monopoly on television patents" (Magoun, 2009, p. 77). The result was a prolonged disinterest in portable civilian cameras, exemplified by RCA Laboratories' "walkie-lookie" being used by NBC reporters at the 1952 and 1956 presidential conventions, "but RCA then abandoned portable, commercial video cameras for twenty years" (Magoun, 2009, pp. xviii, 123).

According to Magoun (2009), in 1968 Sarnoff wrote that by the end of the century, people "equipped with miniature TV transmitter-receivers will communicate with one another" (p. 154). This prediction presents an interesting contrast to Sarnoff's television camera development policies over the preceding forty years.

Magoun (2009) notes a relevant unintended consequence of government regulation in the requirement by the Federal Communications Commission (FCC) that cable television services provide access for locally produced video as a public service (pp. 116-117). This public access, as intended, gave independent and community video producers a wider audience, but also gave those producers access to video equipment they could not otherwise have afforded. These 'amateurs' and their demands became one of the unforeseen drivers of consumer video camera technology.

In his examination of digital imaging technology, Magoun (2009) cites the development of the first live Internet camera at Cambridge, the social phenomenon initiated by the JenniCam webcasts, and the Apple iPhone. Magoun (2009) argues that "Digital internetworks had the effect of removing technical, commercial, or cultural guardians from the medium" (p. 177). As an example, he describes Jennifer Ringley's JenniCam, which consisted of webcasts documenting her life with live video cameras, eventually expanding from one monochrome to four color webcams and remaining online from 1996 until 2003 (Magoun, 2009, pp. 176-177). Ringley's self-revelation to any Internet user was a new one-to-many visual communication form incorporating audience feedback, and has influenced the development of reality TV shows, live web porn sites, and an entire webcam subculture that has become increasingly identified with mainstream society rather than an aberrant subset. Magoun (2009) also points out that Apple's 2007 iPhone combined a cell phone with a camera, LCD touchscreen, keyboard, and Internet access (p. xviii), and observes that by this time, "People had turned cell phones into virtual TV stations, recording video and playing it back or uploading it" (Magoun, 2009, p. 177).

Like Magoun, Abramson (1987, 2003) presents an excellent scholarly resource on the history of television technology. However, the wealth of detail in these two volumes comes at the expense of analysis. The primary utility of these books for the present research is to confirm or to supplement facts presented in the arguments of social historians. Abramson (2003) explicates the 1970s evolution of the professional Electronic News Gathering (ENG) camera that marked the transition from the era of "film at 11" to the immediacy of video news reporting. The 20-pound 'Portapak' ENG was designed for broadcast quality recording, durability, and professional operation, but the author notes that price tags in five figures kept ENG cameras firmly in the category of expensive professional tools (Abramson, 2003, pp. 161-172). Finally, Abramson (2003) argues that development for consumer camcorders shifted to compatibility with the VHS standard because the large numbers of those VCRs in American homes increased the appeal of cameras that could connect to those recorders (pp. 169-171).

Johnstone (1999) writes as a journalist of science and technology, with the primary goal to illuminate the question of "why inventions flourish in one place and not in another" (p. xxiii), making this work valuable to the present research for its wealth of information on technologies that became part of today's complex systems for creating and distributing user-generated online video. Several American innovators lay claim to the charge-coupled device (CCD), the first truly solid-state image sensor, which was the key technology in making consumer electronic imaging practical (Johnstone, 1999, pp. 178-184). Semiconductor memory chips were another American innovation, as was flash memory (Johnstone, 1999, pp. 309, 365, 367).

In a short but densely referenced article, Hintz (2009) examines the development of several types of batteries in relation to the emergence of portable electronic equipment. Hintz (2009) posits that "even before the appearance of the transistor, portable electronic power lagged behind portable electronic equipment and was arguably the more intractable problem" (p. 37). This work is relevant to the present research because the developments the author examines translated directly into improvements in the cost, functionality, and ease of use of portable electronic imaging systems. Today, miniature batteries remain an enabling technology of smartphones, digital cameras, and camcorders alike. Hintz' arguments remain applicable to electronic imaging systems today: battery life is customarily part of reviews of portable consumer electronics, and is often a marketing point as well.

Like Johnstone, Smith (1998) is a contemporary reporter on technology rather than a historian, and his work is therefore weighted toward details of commercial interest with relatively little historical analysis. However, his work is useful to the present research as a broad, inclusive snapshot of the state of electronic imaging technology in

1998, as a very brief overview of the technology's history, as a finding aid to people and companies in the field, and as a source for primary quotations from key actors. In particular, this article addresses the effects of Eric R. Fossum's innovation of the CMOS-APS imager (Smith, 1998, p. 98), which is the core technology of most consumer electronic cameras today.

Smith (1998) observes that the lower power draw and cheaper manufacture of CMOS-APS imagers had enabled building the sensors into a wide range of industrial, medical, and consumer products, beginning the wave of ubiquity of digital cameras (p. 98). In particular, the author's collection of actual and projected sales figures documents both the state of democratization of the technology at that time, and the industry's plans to increase the social permeation of the technology even further (Smith, 1998, pp. 94-100). Presciently, one of Smith's interview subjects remarked that the limitation for the application of CMOS imaging chips in cell phones was the capacity of cell phone networks to handle video data (Smith, 1998, p. 94). Today, network speeds for video transmission are widely advertised marketing points for cell phone service providers.

In their survey-level text on American technology, Marcus & Segal (1999) provide a necessarily brief overview of a number of technologies relevant to the present research, including television, batteries, videotape recorders, and semiconductors (pp. 280-283, 290-291, 371-372). The authors' chapter-based suggestions for further reading also serve as a finding aid for key sources. Marcus & Segal (1999) cite Moore's Law regarding micro-processor chips, "computing power doubles in capacity every 18 months and the number of transistors increases four-fold" (p. 371). The authors note the long-term trends of increased power and stable pricing in encouraging worldwide chip production. This is significant to several of the technologies crucial to the democratization of electronic imaging systems. Marcus & Segal (1999) also observe the growing ubiquity of microprocessors, arguing that the "chips have been fully integrated into the American scene" (p. 372) in applications including video servers and cell phones, two technologies central to the present research.

Like Abramson, Kattelle (2000) presents a chronological narrative of technological innovations, with relatively little historical analysis. This work is therefore most useful as a supplement to works of social history, supplying technical details, dates, and names absent from more summary sources. However, his analysis of the effects of "film at eleven" processes on television news reporting, and the transferral of those film cameras' features to the first portable video cameras, are directly applicable to the present research (pp. 222-236, 341).

Approaching the same issue from another direction, Barnouw (1990) notes the increase in documentary news production in the late 1960s enabled by the mobility of the 16mm film camera equipped with the innovations of electronically synchronized sound and wireless microphones, and argues that these technologies made filmmakers "free agents" (pp. 288-289). This marked the beginning of growing public expectations for visual journalism, and directly informs the present research.

Boyle (1992) argues that the FCC's public access requirements inadvertently provided the hands-on practice that educated a generation of video producers in how to achieve broadcast-quality video (p. 69). The author further argues that this pool of artists, activists, and others without the means to purchase their own professional cameras then drove a part of the consumer demand for affordable cameras capable of producing broadcast-acceptable recordings (p. 69). This is a significant example of government regulation indirectly driving innovation relevant to the present research.

Similarly, Suptic (2009) presents significant evidence that consumer demand succeeded in reversing the development practices of video camera manufacturers from being driven by the requirements of the professional broadcast studios to those of consumer equipment. He argues, for example, that electronic advances designed to compensate for the poor performance of cheap consumer lenses subsequently enabled the use of cheaper lenses in professional cameras (Suptic, 2009, p. 84).

Levinson (2004), writing about cell phones, argues that one persistent reason for the slow adoption of videotelephony is that most people want to control their visual privacy in the home, and points to the relative success of videoconferencing in public or business settings (pp. 170-175). It is useful to remember that he wrote this prior to the advent of YouTube, observing of cell phones in 2003, "Some models could even shoot and transmit video clips" (Levinson, 2004, p. 172). He also cites the historical and commercial antecedents of videotelephony dating back to Tom Swift and his Photo Telephone (Levinson, 2004, p. 171).

Like Levinson, Galambos & Abrahamson (2002) examine the crucial technology of cell phones prior to the full democratization of the technology. Most significantly for the present research, the authors identify the watershed innovation of third-generation (3G) networks using broadband packet transmission and Internet Protocol (IP) packet switching. This enables cell phones to function as Internet devices on the World Wide Web (Galambos & Abrahamson, 2002, p. 242), including uploading images and video captured by camera phones.

Gye (2008), writing on the social practices of cell phones, argues that camera phones are more than just a camera, that they "are both extending existing personal imaging practices and allowing for the evolution of new kinds of imaging practices (p. 135). She also observes that camera phones, combined with Internet image sharing services such as Flickr and YouTube, are transforming how society views both these cameras and the people who use them, in a parallel to the attitudes surrounding the first instantaneous cameras of the late 19th century: "The distrust of the roaming photographer and his/her panoptic technology is resurfacing today in the current distrust of the mobile camera phone" (p. 142).

Carter (2007) presents a chronology of significant dates for the present research, but the brevity of each entry provides no unique arguments, so this work's utility lies primarily in confirming or supplementing facts presented in the arguments of other scholars (pp. 15-16). Similarly, Wiesenfeld (2001) presents a single useful table of numbers, without further scholarly analysis.

YOUTUBE'S SIGNIFICANCE AND OPERATION

YouTube represents the initial cusp of a paradigm shift in mass communication. Significant characteristics of the historical precedents of chemical photography, newswire photo distribution, broadcast and cable television, digital imaging, and cellular radio communication have been surpassed, significantly altered, and incorporated into the innovation of user-generated online video, of which YouTube is the most prominent example.

Strangelove's (2010) discussion makes obvious that YouTube is highly relevant to the present research. The author's four distinctions argue for the special significance of YouTube for videos of purported police misconduct. First, YouTube empowers the class(es) most likely to experience police misconduct. Second, direct commentary and response is enabled for more members of that same class. Third, the mediation of the message through editors and owners whom the police could influence is simply nonexistent with YouTube; with Google data center locations in at least twelve states and fourteen countries (Pingdom, 2008), the ability to unofficially suppress a particular YouTube video is evidently beyond the jurisdictional reach of almost any police department. Finally, any police attempt to suppress a video will, like astroturf, be vulnerable to exposure and ridicule, as more than one law enforcement agency has already learned when they requested YouTube videos be removed, and Google not only refused, but publicized the requests in its Transparency Report (Google, 2011).

Strangelove (2010), Gillespie (2010), and other scholars have identified YouTube as distinct from previous media, particularly because it has a more dynamic and interactive relationship with its audience than has previously been described in classic mass communication models. The present research does not categorize YouTube as a classic publisher. YouTube is a digital media intermediary that its management describes as a platform. Gillespie notes four applicable definitions for platform, in four areas relevant to YouTube. In computation, a platform is support infrastructure; in architecture, it is a raised level surface on which things can stand. Figuratively, a platform is a position achieved from which further activity can be based; politically, it is an issue endorsed by a political party or candidate (Gillespie, 2010, pp. 349-350). YouTube promotes itself as a platform that is open, neutral, egalitarian and that provides progressive support for activity (Gillespie, 2010, p. 352).

For the typical user, YouTube is a free Internet service that enables users to search for, view, and comment on videos uploaded by other users, both amateur and professional. For users who choose to create their own YouTube channel, the service becomes a means of uploading videos to share privately or publicly (Kelly, 2010c).

Who owns YouTube? YouTube, LLC is a wholly-owned subsidiary of Google, Inc. Google, in turn, is a publicly traded company, but over two-thirds of voting shares are held by the company's founders (Efrati, 2012). Thus, YouTube is relatively less affected by the type of corporate and non-employee shareholder pressure that influences traditional corporate media distributors. In addition, Google's unofficial corporate motto, "Don't be evil" appears to inform most of its ethical decision-making (Kelly, 2010c).

YouTube is funded almost entirely by advertising revenue, creating three constituencies for YouTube: users, advertisers, and professional content producers (Gillespie, 2010, p. 353).

Who uploads videos to YouTube? Anyone with a computer, browser software, and Internet access can watch videos on YouTube, but uploading videos also requires registration. To register, the user enters a valid email account, basic demographic data, and accepts the YouTube Terms of Service (detailed on next page). The uploading process itself can be as few as three steps: from the user's YouTube account, click the Upload link, browse to locate the video file, and click the Upload Video button. The user also has the opportunity to add tags and other information about the video, and to choose privacy and social media sharing settings. At the end of the research period, the technical limitations on uploaded videos were that each be less than 15 minutes running time, and less than two gigabytes in size. Multiple common video file formats are accepted, and the help pages have technical recommendations for the best results. Basically, a child who can read can upload video to YouTube; there is essentially no technical barrier for a competent adult (Kelly, 2010c).

According to a survey conducted in 2009 by the Pew Internet & American Life Project, 14% of Internet users have posted videos online, compared to 8% in 2007 (Purcell, 2010). Furthermore, "women are now just as likely as men to upload and share videos, and social networking sites like Facebook are as popular as video-sharing sites like YouTube as locations for video uploading" (Purcell, 2010, p. 2). Age is clearly a factor: 18% of Internet users under age 50 have uploaded a video, while only 10% older than 50 have done so (Purcell, 2010, p. 7). By October 2013, 26% of Internet users had posted original videos online (Duggan, 2013, p. 2).

There has been debate over the advisability of television news operations posting video to YouTube. One argument in favor is that nearly ten percent of users visiting YouTube's news and politics page click on the News Near You video. Some TV stations have decided to put all their video content on YouTube. Hearst Television reports that they make money off their YouTube channels through YouTube's ad revenue sharing. However, most television news operations remain skeptical at this time; only 325 of Google News' listing of 25,000 news sources have agreed to upload video (Potter, 2010, p. 46).

One notable new development regarding YouTube uploaders is the activity of individualized civic watch, which "reverses the idea of oppressive self-regulation and allows people to monitor elites and organizations in order to make their actions more transparent, fair and accountable" (Hayhtio & Rinne, 2009, p. 841). Using YouTube, "it is possible to offer 'official truth challenging' viewpoints on political campaigns and gain access to more personalized information than ever before, which may even reach a global audience" (Hayhtio & Rinne, 2009, p. 840). Such videos may be propaganda to advance a cause, or may be counter-propaganda to expose the activities of others.

Removing videos from YouTube is entirely at the discretion of YouTube itself. It is a private service, so First Amendment protections are not guaranteed. Law enforcement, politicians, businesses, or individuals must first make a formal complaint to YouTube before a video will be considered for removal. YouTube's Terms of Service and Community Guidelines set plain-English limits on what content is acceptable to post on YouTube. Categories of unacceptable content include sex and nudity, hate speech, shocking and disgusting, dangerous illegal acts, children (sexually suggestive or violent), copyright, privacy, harassment, impersonation, and threats. YouTube does not actually monitor incoming videos for all possible violations; an offending video will generally not be considered for removal until users have flagged it. From that point, the process is described in the YouTube Community Guidelines' section on enforcement:

> YouTube staff review flagged videos…to determine whether they violate our Community Guidelines. When they do, we remove them. Accounts are penalized for Community Guidelines violations and serious or repeated violations can lead to account termination. If your account is terminated, you won't be allowed to create any new accounts. (YouTube, 2010a)

In its Transparency Report for 2011, Google highlighted the issue of police requests for removal of YouTube videos documenting police actions:

> We received a request from a local law enforcement agency to remove YouTube videos of police brutality, which we did not remove. Separately, we received requests from a different local law enforcement agency for removal of videos allegedly defaming law enforcement officials. We did not comply with those requests, which we have categorized in this Report as defamation requests. (Google, 2011)

Since 2007 YouTomb, a research project at MIT (youtomb.mit.edu), has tracked videos removed from YouTube, particularly for cases of mistaken allegations of copyright violation. The most common reason for takedown notices is violation of copyright, specifically the Digital Millennium Copyright Act or DMCA. DMCA takedowns are usually requested by the copyright holders' attorneys. Another reason for video removal is violation of privacy; these complaints result in a notice to the video uploader, with 48 hours to respond or to remove or edit the offending video (YouTube, 2010c).

Who watches YouTube videos? According to the Pew Internet & American Life Project, 69% of Internet users (52% of all US adults) watch or download videos online. "The exploding popularity of video-sharing sites like YouTube...has grown from 33% in December 2006 to 61% in the current survey" (Purcell, 2010, p. 3). Broadband access increases online video's popularity to 75%; among reported online video watchers, 89% have broadband at home. Demographically, the affluent, the more educated, men, and young adults are more likely to watch online video (Purcell, 2010, p. 3). By October 2013, the share of online adults who watch or download videos had grown to 78% (Purcell, 2013).

Who posts to discussion threads on YouTube? The same Terms of Service and Community Guidelines that YouTube applies to uploaded video content also control discussion thread posts. In particular, repeated infractions of the rules can result in the offending user's posts being removed from the thread, their YouTube account being deleted, and the creation of any new YouTube account being denied. However, the rules are interpreted rather liberally; discussion posts can use extremely foul language, as long as they are not threatening to an individual (YouTube, 2010a).

It can be challenging to try to narrow down the reasons why individual users participate in YouTube discussion threads. In a single case study, researchers identified 28 emergent codes to categorize thread posts, ranging from support to criticism to jokes to spam (Kennedy, 2010, p. 230). However, other researchers have identified a smaller number of common motivations for users to contribute in the long term: "The social psychological benefits from gratitude, historical reminders of past behavior, and ranking of one's contributions relative to those of others can significantly increase repeat contributions" (Cheshire & Antin, 2008, p. 705). Similarly, the Pew Internet survey found that about 13% of "video viewers who actively exploit the participatory features of online video – such as rating content, posting feedback or uploading video – make up the motivated minority of the online video audience" (Madden, 2007, p. iii).

One of the consistent motivators to user contribution is that a useful pool of information is being built up, a community resource of some value. Even longtime lurkers may find occasion to contribute a useful tip or supportive comment. As Cheshire & Antin (2008) note, "Extremely low costs of contribution combined with very large networks of distribution facilitate production of online information pools—despite an abundance of free-riding behavior" (p. 705).

CHAPTER 3

Research Plan and Process

Case study research can be based on single or multiple cases. For this research issue, a single case would not be sufficiently robust to be useful. Therefore, a multiple case study of 14 cases is carried out. This methodology follows a replication logic, that is, multiple cases are considered as multiple experiments (Yin, 2003a, p. 47), "either predicting similar results (literal replication) or contrasting results for predictable reasons (theoretical replication)" (Yin, 2003b, p. 5). Consistent findings over multiple cases are considered more robust (Yin, 2003b, p. 110). Ten has been shown to be an acceptably robust number of cases for parallel studies (de Graaf & Huberts, 2008). The remaining four cases were added for contrast.

Case studies can be categorized as either exploratory, descriptive or explanatory (Yin, 2003b, 5). There is sufficient definition of terms, concepts, and issues in closely related research areas that an exploratory case study for this research issue does not seem to be required. However, there is not yet a sufficient body of data to make an explanatory case study feasible at this time.

A descriptive case study "presents a complete description of a phenomenon within its context" (Yin, 2003b, p. 5). The phenomenon of a police-civilian interaction in American public space that has been videorecorded by a civilian presents appropriate material for a descriptive case study. A descriptive multiple-case study presents useful opportunities for both direct replication (similar results) and contrast (differing results for expected reasons) among cases, and for cross-case analysis (Yin, 2003b, 5, 110).

The researcher then generalizes the findings to theory (Yin, 2003a, p. 38), including statements as to the conditions under which a particular phenomenon is likely to be found (a literal replication) or not likely to be found (a theoretical replication) (Yin, 2003a, pp. 47-48).

What are the scope and depth of the event (case) being studied? Where should the description start and end, and what should it include and exclude (Yin, 2003a, 28-33; 2003b, 23)?

The event being studied is a police-civilian interaction in American public space that has been videorecorded by a civilian. The scope in time begins with the beginning of the police-civilian interaction event (not necessarily the beginning of the videorecording) and continues through the outcome of that event, and extends backward in time only as necessary to provide context for the event. The outcome is defined as the final ruling for, or agreement between, parties to the event that precludes further legal action. The geographic scope of the case pool includes the area under rule of US law, including states, commonwealths, territories, and possessions. The temporal scope of the case pool is 2005-2010, beginning from the year the YouTube video sharing service went online.

The depth of the event begins with the most complete, unedited and unmodified available form of the videorecording of the event, and includes all reasonably available documents directly pertinent to the event. Pertinent documents include the original videorecording, official statements, police and court filings, depositions and transcripts, press releases, news coverage, editorials, edited and modified versions of the original videorecording, Internet websites, weblogs, and discussion group threads. The case excludes documents that refer only to the research topic without specifically mentioning the event, the principals in the event, or an occurrence pertinent to the event or the outcome. For example, a document on police misconduct that does not mention the event would not be part of the case, but may be included in the context and precedents. All research materials were culled from public sources.

The data acquired are composed mostly of discrete documents. Document analysis for the 38 relevant variables of interest constitutes the primary method of analysis (Yin, 2003a, 85-89; Wimmer & Dominick, 2000, 117-118; Merriam, 1988, 104-118; Prior, 2003, pp. 145-162; Given, 2008, pp. 230-231). As used by Yin, Prior, Firmin (Given, 2008, pp. 754), Weatherbee (Mills et al., 2010, pp. 247), and in the research reported here, the phrase 'variables of interest' is an umbrella term that addresses the limitations of the more commonly used term 'data' by encompassing data values, attributes, characteristics, and indicators; it also includes inferences, interpretations, characterizations, and conclusions drawn from data. It is similar but not identical to classic natural science definitions of 'variable', and is not comparable to the 'dependent and independent variables' as used in quantitative research.

One of the challenges of this type of study is that variables are likely to outnumber data points (Yin, 2003b, p. 5). This study's 14 cases and 38 variables of interest therefore fit within Yin's methodology. A standard form, applied to all cases, was useful in reducing bias in case selection (Prior, 2003). Following de Graaf & Huberts' (2008) solution for "researchers in multiple case studies fac[ing] immense quantities of data... a 'monster grid'" (p. 642) incorporates a row for each case and a column for each variable, thus presenting in a single document a useful at-a-glance summary of the 532 potential variables of interest. This document is presented in multiple-page format in the Appendix.

In the entry for 'Document Analysis' in the *SAGE Encyclopedia of Qualitative Research Methods*, Prior wrote:

> The standard approach to the analysis of documents focuses primarily on what is contained within them…it is also quite clear that each and every document enters into human activity in a dual relation…as receptacles…and as agents in their own right…documents as agents are always open to manipulation by others. (Given, 2008, p. 230)

Consequently, the corroboration and refutation of data presented as factual is especially important for the analysis of this research, as each case is likely to be a highly adversarial phenomenon with strong motivations for one or more principals to obscure or misrepresent facts. Each document was carefully considered in light of its original purpose and intended audience (Yin, 2003a, p. 87). The notes for each case include a detailed chronology, particularly documenting the contemporaneity, simultaneity, and sequence of the creation of multiple documents at times of peak activity in the case. The result of the analysis is a complete description of each case in narrative form, incorporating references to (and where appropriate, excerpts from) each pertinent document. Finally, cross-case analysis was used to synthesize results from variables of interest that are replicated across multiple cases.

PRELIMINARY CONCEPTS

This study is interdisciplinary, specifically including the analytical tools of mass communication and of history. The context and precedents also include pertinent aspects of media law, history of photographic technology, and history of police-civilian interaction.

The unit of analysis (case) is an event of police-civilian interaction in American public space that has been videorecorded by a civilian, and the sequelae and final outcome of that event. The case selection and screening criteria include: American public space; police-civilian interaction; clear final outcome; user-generated online video (UGOV), that is, video recorded by an unaffiliated civilian rather than a journalist, a government employee, an organization, or a company. At least one case was selected that does not include user-generated online video for contrast (differing results for expected reasons). At least one case has been selected that includes online video generated by traditional mass media, also for contrast.

Thirty-eight relevant variables of interest were identified. Following are explicit and specific rationales for why each variable has been selected and, in particular, why each variable is related to the (possible social, cultural, or technological) historical or legal discussion in this research.

The variable of whether the police were aware of the camera is relevant to this study because most people become more self-conscious and will modify their behavior if they are aware of being recorded by a camera. This variable is therefore crucial to the accurate interpretation of the video evidence. If the police were not aware of the camera, one may reasonably suspect that their actions were less inhibited and more natural. Closely tied to this is the variable of when the police became aware of the camera, and whether there was any marked change in police behavior once they were aware of the camera. These three variables may be deduced from the video evidence, and do not rely on self-reporting or other more subjective measures. These variables are particularly relevant to historical patterns of police cultural attitudes toward civilian cameras, and are also relevant to the technological discussion of camera size, capabilities, and obtrusiveness.

The three interrelated variables of police action regarding videorecording include: whether police made any attempt to prevent the recording; whether police made any attempt to acquire, confiscate, or destroy the video; and whether the police gave any unlawful instruction to the videographer regarding the video. These variables are relevant primarily for legal and for police cultural reasons. Legally, the act of videorecording is highly unlikely to be unlawful due to the scope of the research design. Therefore, any police action or instruction that would prevent, confiscate, or destroy the recording is itself likely to be illegal, or at least an abuse of police discretion. These three variables are also relevant and significant indicators of police cultural practices regarding civilian cameras.

The variable of whether police misconduct was or may have been recorded is relevant to the legal discussion and to the discussion of police cultural practices because the recording may constitute evidence material to an investigation. Police actions to preserve, control, or destroy evidence must be evaluated differently from police actions in response to nonevidentiary recording.

The variable of whether police detained, cited, or arrested the videographer is relevant to the legal discussion and to the discussion of police cultural practices. Police guidelines and court rulings for search & seizure are specific as to permitted actions with a detained person as opposed to an arrested person, including the inspection or confiscation of cameras and recording media. Detention, citation, and arrest are also historically part of police cultural practices in punishing persons who are not in prosecutable violation of law, but who have annoyed the police, e.g., videorecorded police actions. The combination of this variable with the subsequent dropping of charges may constitute evidence that the initial police action was intended to be punitive in itself.

The variables of when the video was available via the Internet, when the video was available via broadcast news media, and when images from the video were available via print news media are relevant to the technological historical discussion of visual media distribution technologies. The date and time for each of these variables also marks the beginning of public discussion of the respective publication.

The variable of whether more than one camera captured the event is relevant to the evaluation of the veracity and comprehensiveness of the video evidence; the more points of view are available, the more complete, reliable, and legally definitive the video evidence is likely to be.

The variable of whether official CCTV, dashboard, or other video captured the event is relevant to the legal discussion, and to the technological and police cultural historical discussions. Official video recordings may have greater credibility than civilian video due to certified time and date stamping and a procedural evidence trail. Police organizations have historically been early adopters in visual recording for specific purposes. Police cultural practices have also been observed to change in the presence of official cameras.

The variable of whether police initially admitted to the existence or possession of video of the event is relevant to the discussion of historical police cultural practices. Similarly, the variables of when did police admit to possession of video of the event, whether police released official video of the event, and whether official video of the event was available via the Internet are all relevant to the discussion of police cultural practices, particularly any tendency of police to suppress evidence of police misconduct. These variables may also be relevant to the legal discussion, as some courts have ruled that official videos are public documents within the meaning of the Freedom of Information Act.

The variable of who was credited as the source for each medium of release of the video is relevant to the technological historical discussion because news media standards and practices are evolving in response to technological innovations such as YouTube. This variable is relevant to the legal discussion because legal issues of copyright and fair use are also evolving in response to technological innovations such as YouTube.

The variable of whether there was any effort to restrict, remove or prosecute the release of the video (takedown order, SLAPP, etc.) is relevant to the discussion of historical police cultural practices, and to the legal discussion. Because the scope of the research is carefully constructed to preclude legitimate takedown grounds of privacy or copyright, any such effort will most likely be unlawful, and may provide evidence of misconduct.

The four variables of what was the initial response of the law enforcement agency regarding the officer(s) involved, what was the final outcome regarding the officers involved, what was the sequence of official actions regarding the officers involved, and what was the final outcome of the law enforcement agency regarding policy changes, if any, are all relevant to the legal discussion and to the discussion of historical police cultural practices. Legally, the law enforcement agency is bound by law and by written administrative procedures. Culturally, the police who make up the agency often have unofficial and unwritten practices with long historical precedents. A careful examination of these four variables may enable the deduction of unwritten practices by 'reading between the lines' of the agency's legal and administrative actions.

The five variables of the videographer, including: the initial official response to handling of the videographer; the final outcome regarding the videographer; the sequence of official actions regarding the videographer; the compensation (if any) of the videographer; and the affiliation of the videographer; are all relevant to the technological historical discussion, the social historical discussion, the police cultural discussion, and the legal discussion. Videographer capabilities are the result of historical technological developments. What happens to the videographer, in both the short and long run, has significant implications for the practice of public space photography. Police actions toward the videographer, and the outcomes of those actions, may influence police cultural practices. The affiliation (or lack of it) of the videographer may influence initial police actions, police organizational responses, the credibility of the video, and the drawing in of third parties such as the National Press Photographers Association or American Civil Liberties Union.

The five variables of the subject, including: the initial official response to handling of the civilian subject, if any; the final outcome regarding the subject; the sequence of official actions regarding the subject; the compensation (if any) of the subject; and the affiliation of the subject are all relevant to the social historical discussion, the police cultural discussion, and the legal discussion. Police actions toward the subject, and the outcomes of those actions, may influence police cultural practices. The affiliation (or lack of it) of the subject may influence initial police actions, police organizational responses, the credibility of the video, and the drawing in of third parties such as the ACLU.

The three variables of how much press coverage the event received, how much Internet discussion was linked to the event, and how many times the video was viewed online are relevant to the technological historical discussion, the social historical discussion, and the police cultural discussion. Press coverage has historically been one measure of the societal permeation of a story, and as interpreted through the agenda-setting model, it can have a significant effect on political matters that affect both police and civilians. Video viewing tallies are a technologically new measure that may be comparable to press coverage. Internet discussions are a technologically new form of more-or-less public interaction that may prove to be significant to social, cultural, and legal development in ways similar to the historical influences of newspaper op-ed pieces, radio call-in shows, and television viewer participation programs.

The variable of what third parties (if any) involved themselves in the legal case is relevant to the legal discussion, the police cultural practices discussion, and the technological historical discussion. Involvement of pro-police organizations such as the FOP may be a significant indicator of police cultural practices. Involvement of civil rights organizations such as the ACLU may be a significant indicator of both the type of issues presented by the case and of the relative importance or severity of the case. Involvement of organizations such as the Electronic Frontier Foundation or NPPA may indicate legal issues related to new technologies.

The variable of *cui bono?*, or who benefits?, is relevant to the legal discussion and to the social, cultural, and technological historical discussions. Any court ruling conferring a benefit on one or the other party in the case would plausibly encourage similar parties in other cases to pursue similar legal strategies; a significant monetary award would likely be followed by similar lawsuits. On the other hand, rulings that validate police actions and assess court costs to civilian plaintiffs could be expected to chill prospects for the filing of similar suits. Historically, outcomes perceived as unjust by a large fraction of American society have led to social, cultural, and legal changes; an apparent benefit to a malefactor might be expected to instigate such changes. Technologically, a material benefit to the party supported by the video evidence might lead to increased sales, further development, and further societal permeation of the recording technology. On the other hand, a negative outcome for technology users might lead to restrictive legislation, adverse court rulings, or changes in social or cultural practices in order to discourage use of the technology.

Data collected include: the event's most complete, unedited and unmodified videorecording that is reasonably available is the primary document, followed by the official police report of the event. Internet data collected include: the video sharing site statistics and related tags, responses, and threads about the event; weblogs citing the event; discussion group threads citing the event; and traditional news media web pages reporting on the event. Broadcast and print media reports on the event were also collected. Public statements and press releases from any of the principals or involved third parties were likewise collected. Legal documents, if any, were collected. Date and source information for each document represents crucial data in building a chronology.

DOCUMENT SEARCH AND ACQUISITION PROCEDURES

As Prior recommends, "the reasons for including and excluding cases ought to be defined in advance of any study" (2003, p. 150). The primary means of identifying potential cases for this research has been to personally monitor the mass media and relevant legal and academic publications, mostly through the Internet, and to note the principals and basic information for each potential case. This identification process ran continuously from July 2008 through 2011. This process was not exhaustive, but was adequate to identify a sufficient number of cases meeting the selection criteria to provide a large enough pool of cases to yield acceptably robust findings. The selection criteria of public space and of user-generated online video were sufficient to rule out a large number of police-civilian interactions that were recorded on official or professional video, or that occurred in private space. Of the remainder, a relatively small subset presented any public documents beyond the original video; in particular, a very small number presented public documentation of a clear outcome. This subset represents the ten cases originally selected for direct replication (similar results for expected reasons), per Yin (2003, pp. 5, 110). Ten has been shown to be an acceptably robust number of cases for parallel studies (de Graaf & Huberts, 2008). The remaining four cases were added for contrast (differing results for expected reasons), including the absence of video, the professional status of the videographer, and a video that documents correct police conduct.

The second stage of the data gathering entailed collecting all available relevant documents regarding each case. The first difficulty with research based on Internet content is that the data are mutable at best, and often ephemeral. For reliable analysis, it is useful to duplicate the online data into a local, fixed, and searchable form that is under the control of the researcher, yet preserves the look and feel of the original online interface. Different document types require different tools and procedures for search, acquisition, and review.

For each video recording, the procedure was to: 1) follow up any links provided by traditional mass media, blogs, or discussion groups; 2) identify duplicates or alternative forms of the video; 3) select (when possible) the oldest, the longest, the least edited, and the highest quality versions of the video; 4) use tools such as aTube Catcher (Uscanga, 2010) to create a local copy of the online video; and 5) note all available data about the site, source, and the video itself.

For video sharing site statistics, tags, responses, and threads, the procedure was to: 1) search the site, using the internal search function, for duplicate or alternative forms of the video; 2) Save the selected video page as a complete archive, including both HTML code and embedded content; 3) follow up links to additional comments and responses; 4) archive pages containing relevant comments or responses; and, 5) note dates of retrieval of all files, preferably in the file or folder name.

For legal documents, the procedure was to: 1) search LexisNexis Academic for the names of the principals of the selected case studies; 2) identify the correct case citations; and 3) download copies of all related case filings, in particular the final rulings. These documents are available in HTML, doc, and PDF formats. For more recent filings not yet in Lexis, a search of Justia.com was carried out. For legal filings from federal appellate, district and bankruptcy courts, a PACER (Public Access to Court Electronic Records) account was set up. For legal documents not available online, the clerk of the court was contacted to secure the relevant documents.

For police reports, the procedure was to: 1) search the Internet using Google and other search engines for the names of the principals and terms such as 'arrest report' combined with file formats such as PDF, JPG, and PNG; or 2) search for the law firm or ACLU office representing the plaintiff, and examine their website for posted documents; or 3) search the website of the police department or municipal or county government having jurisdiction, or 4) contact the police department or clerk of the court and pay for duplication and postage of the relevant documents.

For traditional news media web pages, the procedure was to: 1) search the site, using the internal search function, for duplicate or alternative forms of the story and for any mention of the names of the principals; 2) Save the selected page as a complete archive, including both HTML code and embedded content; 3) follow up links to additional stories, comments and responses; 4) archive pages containing relevant stories, comments or responses; and, 5) note dates of retrieval of all files, preferably in the file or folder name.

For print media reports, the procedure was to: 1) search the wire services for reports including the names of the principals; 2) search Internet news aggregators for reports including the names of the principals that are not derived from the wire services; 3) search Access Newspaper Archive, ProQuest Newspapers, and other databases including newspaper and news magazine content. For any relevant reports, a local copy was saved with appropriate date and source data included.

For broadcast media reports, the procedure was to: 1) search the websites of the broadcast stations in the area where the incident occurred; 2) search the websites of the national broadcast networks; 3) archive relevant reports, including embedded content and the appropriate date and source data.

For public statements and press releases, the procedure was to: 1) gather the names of spokespersons and the dates of statements from other media reports; 2) search the wire services for reports including the names of the spokespersons; 3) search Internet news aggregators for reports including the names of the spokespersons that are not derived from the wire services; 4) search Access Newspaper Archive, ProQuest Newspapers, and other databases including newspaper and news magazine content; 5) retrieve a complete copy of the prepared statement or press release, with appropriate date and source data; and, 6) if the statement was on camera, collect a copy of the video using the appropriate procedures.

For discussion group threads, the procedure was to: 1) identify a discussion group by searching for the names of the principals of the selected case studies; 2) confirm the relevance of the discussion group by reading a sample of threads; and then 3) make a snapshot of the discussion group. As an example, I initially identified the Flickr

"Photography is Not a Crime" discussion group (hereafter PINAC) through multiple links and references in source material from the preliminary case study selection. I selected PINAC based on the span of time it has existed, the number of threads or discussion topics, the apparent high level of activity among the core members, and the initial appearance of content of interest to the present research. PINAC has a large but not unwieldy membership, over two years of activity, and a clearly stated official focus. I used the program HTTrack (Roche, 2010) to download a snapshot image of PINAC's discussion forum, www.flickr.com\groups\photography_is_not_a_crime\discuss, on March 29, 2010, beginning at 4:34 p.m. and completing the download at 6:44 p.m. This snapshot data set is searchable with a variety of tools, and remains browsable in the same way as the original online data. The data set for the discuss folder itself totaled 9,094 files in 9,089 sub-folders, occupying 38.6 megabytes of storage. There were also external links to 55 other Internet domains, whose folders contained 808 files occupying 68.6 megabytes of storage. Some basic statistics about PINAC may serve as a useful background.

At the time of the download, PINAC's forum showed an index of 31 pages of threads, with a total count of 613 threads. Reported membership was 2,138 with six group administrators. I imported the thread index pages to an Excel spreadsheet, which calculated an average of 13.7 replies per thread overall, increasing to 17.5 in the most recent six months and 18.8 in the most recent three months. There were 50 posts with no replies; these were almost all links to external resources. There were five threads with 100 or more replies; the longest thread had 146 replies. There were 8,435 replies to the 613 threads, for a total of 9,048 individual messages posted during the period from March 19, 2008 to March 29, 2010.

For weblogs, the procedure was to: 1) identify a blog by searching for the names of the principals of the selected case studies; 2) confirm the relevance of the blog by reading a sample of posts; then, 3) make a snapshot of the blog. This is essentially the same procedure followed for discussion groups.

GENERALIZING TO THEORY

The goal of the present research was to generalize from the multiple-case study to useful theory (Yin, 2003a, 38). Examination of the relevant variables of interest within and across cases was used to identify common elements of the examined events which are likely to recur in future similar events, common elements which present important implications for freedom of the press and other civil liberties, and broader theoretical issues that merit further study.

The research question and the hypotheses for this research were:

RQ1: What is the outcome of user-generated online video on police-civilian interactions in American public space?

H1: User-generated online video has the potential to improve accountability in police-civilian interactions in American public space.

H2: User-generated online video is significantly different from professional video journalism in its effects on accountability in police-civilian interactions in American public space.

H3: There are strong motivations for police to continue to attempt to restrict civilians' First Amendment rights to photograph police in public spaces.

It is customary and useful to cite previous scholarship that bears methodological resemblance to new research intended to develop theory. In the present case, that citation has proven impractical due to the almost complete lack of comparable research. Several factors contribute to this relative scarcity. First, online video sharing sites are historically very new phenomena; YouTube only went online in 2005. Second, most case study research is in areas that favor personal interviews as primary data, so document analysis is an uncommon methodology. Third, multiple case studies with cross-case analysis are generally used to examine and compare organizations, so research applying this methodology to individuals or incidents is also relatively uncommon. Nevertheless, a few publications do bear some relevance to the present research. The following four publications are presented in order of relevance.

de Graaf & Huberts, 2008, Portraying the Nature of Corruption Using an Explorative Case Study Design, in *Public Administration Review, 68*(4), 640-653. The authors attempted to discover the nature of corruption within the scope of Western democracies. They applied an explorative multiple case study methodology to analyze the contents of confidential criminal case files, and a number of supplemental interviews with case investigators. Generalizing to theory, de Graaf & Huberts offered nine 'propositions' to describe a general profile of corruption.

This is the closest example, methodologically, to the research reported here; it is a multiple case study with cross-case analysis. Ten cases are selected. The research design is taken directly from Yin on most points. Each case examines the misconduct of one individual in his or her official capacity. The data are culled primarily from documents, which include recorded conversations, media reports, and court verdicts. The characterization of the case studies as explorative is not precisely correct, as the researchers also extended the case studies into the descriptive realm, which makes it even more similar to the present research. One difference is that the researchers also conducted 15 new interviews with the case detectives and their superiors. Other differences are that the research is not geographically limited to the U.S., and that it is working within a different legal system.

Pantti & Wahl-Jorgensen, 2007, Disaster Journalism as Therapy News? The Political Possibilities of the Spectacle of Suffering, presented at the 2007 Annual Meeting of the International Communication Association. The authors studied six British national disasters as reported in two national newspapers, using discourse analysis to examine the portrayal of ordinary people, governmental accountability, and the moralizing of journalists. The authors found that the disaster coverage discourses included horror, anger, empathy, and grief.

This paper parallels the methodology of data collection from published media reports, defining each case as a specific incident, and of using multiple case studies. However, it uses discourse analysis rather than document analysis within the individual cases. This paper's scope is also historically broader and geographically outside the U.S., and there is no evidence in the form of audiovisual recordings.

Goldsmith, 2010, Policing's New Visibility, in the *British Journal of Criminology, 50*(5), 914-934. This paper uses two incidents of apparent police misconduct, the Tomlinson case in 2009 and the Dziekanski case in 2007, to support Thompson's (2005) theory of 'new visibility' regarding mobile phone cameras, YouTube, Facebook, and police image management.

This paper is similar in topic to the present research, and it uses multiple case studies. Data are culled from media reports, official documents, and YouTube and other online video sites. However, this paper uses the two case studies to illustrate a theory, rather than being purely descriptive.

Karpf, 2010, Macaca Moments Reconsidered: Electoral Panopticon or Netroots Mobilization?, in the *Journal of Information Technology & Politics,* 7(2/3), 143. The author argues that the influence of YouTube on politics is often overstated, and that the context of the campaigns and organizations must be taken into account. To support this argument, the article examines two high-profile candidate gaffes through the archived blog entries on DailyKos.

This article is only tangentially similar to the present research, in that it uses YouTube and a blog as data sources, and it constructs and compares two case studies. However, it uses the case studies to support an argument rather than being purely descriptive.

CHAPTER 4

Case Studies

CASE STUDY I: SANTO, BUSH, RIVIERI, 2007-07-01

On July 1, 2007, fifteen-year-old Tony Santo videorecorded Baltimore Police Officer Salvatore Rivieri verbally and physically abusing fourteen-year-old Eric Bush in a public plaza of Baltimore's Inner Harbor. This case is included for direct replication (similar results for expected reasons), per Yin (2003, pp. 5, 110); the phenomenon is a police-civilian interaction in American public space during which police misconduct may have occurred, and of which user-generated video was posted online.

Research Question and Hypotheses

For Research Question 1, what is the outcome of user-generated online video on police-civilian interactions in American public space?, the outcomes of this case include: suspension, reassignment, and eventual firing of the police officer, with loss of pension (Hermann, 2010b); a new lieutenant and sergeant took command of the twelve officers in charge of patrolling the area (Linskey, 2008); the department redesigned the complaint-handling portion of its website (Baltimore City Police, 2008); the department stopped disclosing the names of police officers who shoot or kill citizens (Hermann, 2009a); the Bush lawsuit was dismissed for failure to file before the deadline, with costs to the plaintiffs (Augenstein, 2009); and the videographer attained some notoriety (Santo, 2008).

For this case, Hypothesis 1, that user-generated online video has the potential to improve accountability in police-civilian interactions in American public space, appears to be supported. Two examples of user-generated online video (Santo, 2008; Exzellent, 2008) documented a police pattern of misconduct, and thus improved police accountability by setting the public agenda such that traditional police organization attempts to shield the police officer from the consequences of his actions were not successful. The officer was fired, the supervisor was reassigned, and several internal police policies were revised. An ongoing pattern of police suppression of civilian complaints was also revealed during this case (Chapman, 2008; AmateursGuide, 2008; WJZ, 2008; Jimster1956, 2009), and the police department issued public statements announcing reforms to the complaint process (Baltimore City Police, 2008). This case also demonstrated accountability for the civilian in the interaction in that he admitted he was wrong (Jones & Sentementes, 2008). He reportedly no longer skates in public spaces, and his (and his mother's) lawsuit was dismissed with court costs to plaintiffs (Hermann, 2009c).

The value of Hypothesis 2, that user-generated online video is significantly different from professional video journalism in its effects on accountability in police-civilian interactions in American public space, in this case does not appear to be evaluable without conjecture.

Hypothesis 3, that there are strong motivations for police to continue to attempt to restrict civilians' First Amendment rights to photograph police in public spaces, appears to be supported in this case. The video evidence provided a clear enough picture of the police actions to enable other police to judge accurately (as several did) that while the outline of the police response was appropriate, the manner used was "over the top" and constituted misconduct (Weinblatt, 2008; Officer.com, 2010). This case thus provides an example of a police officer losing his job and his pension because his actions were videorecorded, and his organizations (both department and fraternal) lost public respect (Hermann, 2011; Meister, 2010). These losses provide strong motivations for police to prevent this from happening to themselves or to their organizations.

Documents

The case data exceed 300 files, including 29 media clips from YouTube, broadcast TV, radio, and BPD official releases. The court documents are few but crucial. The *Baltimore Sun* is a major news source, particularly the Crime Beat reporting of Peter Hermann.

Tony Santo's video was posted to YouTube February 9, 2008, and remains online today (Santo, 2008). Baltimore Police Department regulations require that police issue a Citizen Contact receipt to the citizen and file a report for incidents like this one. Rivieri did neither, leaving no official paper trail for the incident. At another time in the Inner Harbor that summer, art student Billy Friebele also had an interaction with Officer Rivieri which was recorded as part of the video component of the art project Friebele was performing (tesla121, 2008). Again, Rivieri failed to issue a Citizen Contact receipt to the citizen, or to file a report.

Bush and Santo both gave interviews that appeared in print and on television (Good Morning America, 2008; WJZ, 2008; WMAR, 2008; Jones & Sentementes, 2008). Bush's mother stated in a televised interview that she had attempted to file a complaint (Good Morning America, 2008); a BPD spokesman stated that Eric never made an official complaint and that Rivieri had no other citizen complaints in his file (Linskey & Sentementes, 2008). Forum, comment and blog posts from at least three credible civilians documented efforts to file complaints against Rivieri, which BPD officers refused to process (Chapman, 2008; AmateursGuide, 2008; WJZ, 2008; Jimster1956, 2009). Both Roy Bush, Eric's father, and Rivieri aired case details on call-in radio programs (Ehrlich, 2010; WBAL & Lang, 2010a, 2010b).

Internal BPD documents introduced in court show that the internal investigation charged Rivieri with using excessive and unnecessary force and "discourtesies" (Hermann, 2010f). Statements by police sources attest that the commissioner offered Rivieri a 90-day suspension and anger management classes, which he refused, and that he insisted on a review board (Hermann, 2010f). The report of the review board found that Rivieri had failed to file required paperwork, and recommended less than a week's suspension (Hermann, 2010d). The police commissioner, based on the board's findings, fired Rivieri (Hermann, 2010b). The appeals court upheld the firing (Hermann, 2011). Court documents show that Miller filed suit seeking $6 million for assault, battery and violation of rights, but filed late, so the case was dismissed (*Miller v. Rivieri*, 2008).

Relevant Variables of Interest

The variable of **whether the police were aware of the camera** is positive. This can be deduced from the video (Santo, 2008), in which Rivieri appears unaware of the camera until 03:35:00, which is the variable of **when the police became aware of the camera**. At that time, Rivieri appears to react to the camera by redirecting his attention from Bush to Santo, asks Santo if the camera is on, and immediately issues a thinly veiled threat; this is the variable of **whether there was any marked change in police behavior once they were aware of the camera**. According to Santo, he held the camera near his thigh so Rivieri would not realize he was being recorded; "I was like, I can't believe I'm getting this on film" (Jones & Sentementes, 2008). As soon as Rivieri asked if the camera was on, Santo turned it off because, he said, "I didn't want to tell it's on. He's already pushed [Bush] down, I'm afraid he's going to take the camera. He knows he did something wrong. If he didn't do something wrong, he wouldn't be asking about the camera" (Jones & Sentementes, 2008).

The variable of **whether police made any attempt to prevent the recording** is positive. Rivieri made an attempt to prevent further recording by his threat at 03:37:20 (Santo, 2008). The variable of **whether police made any attempt to acquire, confiscate, or destroy the video** is negative. There is no evidence that Rivieri made any attempt to acquire, confiscate, or destroy the video, other than by his threat at 03:37:20 (Santo, 2008).

The variable of **whether the police gave any unlawful instruction to the videographer regarding the video** is positive. Rivieri's coercion of Santo to cease recording was an attempt to suppress evidence of police misconduct (Santo, 2008, 03:37:20). The variable of **whether police detained, cited, or arrested the videographer** appears to be negative. There is no evidence that Rivieri detained, cited, or arrested Santo beyond the temporary detention of all the teenagers while Rivieri confronted Bush. Threats of arrest recorded in the video appear to be directed only at Bush. None of the teenagers were cited, as evidenced by Rivieri being charged with not filing mandatory paperwork for this kind of incident (Santo, 2008; WBAL & Lang, 2010b).

The variable of **when the video was available via the Internet** is documented on camb0i's YouTube channel as Saturday, February 9, 2008 (Santo, 2010). The variable of **when the video was available via broadcast news media** is reported as being via WJZ-TV late Saturday, February 9, or early Sunday, February 10, 2008 (WJZ, 2008). The variable of **when images from the video were available via print news media** is on page A1 of the *Baltimore Sun* on February 12, 2008 (Linskey & Sentementes, 2008).

The variable of **whether more than one camera captured the event** is negative, based on the document search and acquisition procedures. All publicly reported evidence appears to indicate that only Santo's camera captured the event.

The variable of **whether official CCTV, dashboard, or other video captured the event** is apparently negative. Baltimore's Inner Harbor area has been covered by a network of police surveillance cameras since 2005 (Janis, 2005). No mention of these cameras was made in this case. Thus, the variable of **whether police initially admitted to the existence or possession of video of the event** is negative. The related variables of **when did police admit to possession of video of the event**, **whether police released official video of the event**, and **whether official video of the event was available via the Internet** are therefore not applicable.

The variable of **who was credited as the source for each medium of release of the video** is that Tony Santo has evidently never been credited by name as the source for any medium of release of the video. His camb0i YouTube channel does not include his real name, and the broadcast television stations that used his video either omitted onscreen credit entirely, superimposed their own station logo, or credited YouTube (bbbbill123, 2008; WBAL, 2010b, 2010c; WMAR, 2010).

The variable of **whether there was any effort to restrict, remove or prosecute the release of the video** is apparently negative. It is not evident from the available public documents whether there was any discussion of this option for Rivieri or the BPD.

The variable of the **initial official response to handling of the videographer** is that there was no initial official response to his handling (Jones & Sentementes, 2008). The variable of the **sequence of official actions regarding the videographer** is that the only official action was a single interview with IID detectives (Jones & Sentementes, 2008). The variable of the **final outcome regarding the videographer** is not applicable because there was no official response to police handling of the videographer. The variable of the **compensation of the videographer** is unknown.

The variable of the **affiliation of the videographer** is that he was a civilian and was not affiliated with any media or organization other than his high school. However, Bush admitted he and his friends have often interacted with police because of their skateboarding, and Santo stated he recorded the interactions because the teenagers think it is funny when police yell at them (Jones & Sentementes, 2008). A skateboarding reporter pointed out that "Kids are tech-savvy, and as important as going out and learning a new trick is filming a new trick" (Emery, 2008). Camera-carrying skateboarders have uploaded an increasing number of user-generated online videos documenting confrontations between skaters and police. Thus, it may be valuable to consider the videographer's affiliation with the skateboarding community.

The variable of **whether police misconduct was recorded** is: the internal review by Baltimore PD Internal Investigation Division detectives, including examination of this video, came to the conclusion that Rivieri did use excessive and unnecessary force and discourtesies (Hermann, 2010f), misconduct for which he could be fired. Thus, police misconduct was recorded by Santo's video.

The variable of the **initial response of the law enforcement agency regarding the officer involved** was when police answered a call from Miller: "I called to file a complaint but was told the supervisor was on vacation and would call me when he returned. He never did" (WJZ, 2008).

The variable of the **sequence of official actions regarding the officer involved** is complex. In simplest form, he was suspended with pay, investigated by IID, returned to patrol duty, investigated by IID again, charged with misconduct, offered a plea deal of a 90-day suspension plus anger management classes, refused the deal in favor of a trial board, was partially cleared by the trial board, and was fired by the commissioner (Hermann, 2011).

Sunday morning, February 10, the Baltimore Police Department received an email from a reporter with the *Baltimore Sun*, including a hyperlink to Santo's video. Police told reporters that they did not know the identities of the boys involved. Police spokesman Sterling Clifford said the police were contacting area schools to try to find the person who uploaded the video (Linskey & Sentementes, 2008).

The BPD Internal Investigations Division began an investigation into the incident on February 11, stating in part, "As a result, Officer Rivieri's police powers were suspended" (Skinner, 2008). Clifford stated to Sher, "The police commissioner and the mayor and the top command staff saw the video this morning and are disappointed. The officer has been suspended with pay, pending an internal investigation" (WJZ, 2008). Clifford also said that the suspension means Rivieri has been transferred to administrative duties, that the boy never made an official complaint, and that Rivieri's file contains no other citizen complaints (Linskey & Sentementes, 2008). Baltimore Mayor Sheila Dixon stated, "I saw the tape. The officer handled the situation so poorly. It is very clear this is unacceptable" (WJZ, 2008).

Tuesday, February 12, Clifford responded to queries from multiple reporters with statements including, "The entire incident raised red flags for all of the members of the command staff who watched the video," and "We have invested a lot of time and energy in having better relations between the community and the police. The bad behavior of one police officer can jeopardize a lot of hard work" (Linskey & Sentementes, 2008).

Police reportedly had refused to accept complaints against Rivieri from three other civilians. When they went to one location to file a complaint, police told them they had to go to a second location; when they tried the second location, police directed them back to the first location. It was made clear that police were not going to accept their complaints (Chapman, 2008). After viewing the Santo video, civilian Jon Tarburton recognized Rivieri and filed a complaint with IID on February 11 (Jones & Sentementes, 2008). By February 14, the department was reportedly looking into Tarburton's complaint; Clifford confirmed that a second complaint had been filed against Rivieri since the YouTube video surfaced. The same day, IID detectives interviewed Santo, with his mother present, for details about the context of the video (Jones & Sentementes, 2008).

Police investigators asked another civilian to make a formal complaint against Rivieri after the civilian posted to a forum thread about a similar incident. The civilian did so, speaking with two detectives, who also asked for a link to the thread (AmateursGuide, 2008b).

On February 16, Clifford commented on the Friebele video that "this second video may be evidence to be included" in IID's investigation of Rivieri (Emery, 2008).

On April 22, the department made significant changes to police patrols of the Inner Harbor. The patrol command changed to Lt. Samuel Hood III, formerly a supervisor in planning and research, and Sergeant Henry Wagstaff, formerly of special operations. Two patrol officers were added, for a total of twelve; more vehicles were added to raise the visual profile of the unit. All unit officers attended sensitivity training. Clifford explained, "Given the extreme nature of that incident, we thought it was important for the officers to brush up on their interpersonal skills" (Linskey, 2008). Clifford also stated that Rivieri was still on administrative duty, and that the IID investigation was continuing (Linskey, 2008).

On the first of May, 2008, the Baltimore City Police website page for "How do I file a complaint about an officer?" was updated. The text addresses a number of the criticisms and concerns expressed in public forums following Rivieri's suspension. The middle section of the web page provides information for additional organizations, including addresses and telephone numbers for the Human Relations Commission, Community Relations Commission, and Legal Aid Bureau. The final section of the web page is a warning that the process is not for frivolous or false accusations (Baltimore City Police, 2008).

July 14, 2008, Baltimore Police Department Associate Legal Counsel Michael F. Conti, representing Rivieri, asked the court to dismiss Miller's civil case (*Miller v. Rivieri*, 2008).

In October, IID completed its investigation. On October 24, Major Tiedemann reported the investigation complete with no criminal or termination findings. Five days later, Rivieri was medically evaluated for fitness for duty at Mercy PSI. On November 5, Rivieri was medically released for full duty. At Rivieri's two-minute-long suspension hearing on the seventh, Colonel John P. Skinner signed and dated the internal memo for Rivieri's restoration to full duty (Skinner, 2008). Rivieri, his defense counsel Herbert Weiner, the Inner Harbor unit commanding officer Lt. Hood, and secretary Denise McNeill were present at the hearing. Weiner asked that his client be reassigned to the

Southeastern District, which Skinner approved. By November 10, 2008, Rivieri was back on patrol duty (Skinner, 2008).

After Rivieri's return to patrol duty, a new internal investigation was launched, the results of which would not become public for nearly two years (Hermann, 2010f).

On December 12, police spokesman Donnie Moses made statements to the media that appear to contradict Rivieri's suspension hearing memo. According to Moses, Rivieri "was charged departmentally, relieved from street duty, inside doing admin duties," was still suspended, and would remain so until the civil suit and internal trial board proceedings were completed (Kearney, 2008). The wire service version of the story also stated, "The police department says Rivieri is still relieved from street duty" (Linskey & Sentementes, 2008). At the time of Moses' statement, Rivieri had been back on the street for over a month.

On January 7, 2009, the department changed its long-standing policy and would no longer release the names of police who kill or injure people (Fenton, 2009). Department spokesman Anthony Guglielmi referred to Rivieri's case, saying that Rivieri had received death threats at his home (Fenton, 2009). Notably, Rivieri's home address and phone number had been posted online over 150 times in comments about the YouTube video (Santo, 2008). The department changes affecting transparency elicited press complaints:

> Police say they will release the name of the officer if the department rules the shooting unjustified. If it's justified, apparrently they won't release anything...the department doesn't release the outcome of any internal investigation...regardless of how it turns out...trial boards are almost impossible to attend because the city refuses to give us a schedule. (Hermann, 2009a)

On September 14, 2009, the court granted Rivieri's defense motion for summary judgment in the civil suit, and awarded costs to the plaintiffs, Bush and his mother (*Miller v. Rivieri*, 2008). City Solicitor George A. Nilson said, "Deadlines are deadlines, and good cause is more than just, 'I forgot' or 'I got confused'" (Kearney, 2009). Nilson also said that he does not think an apology from Rivieri is in order. In the same story, Guglielmi said an internal investigation against Rivieri based on a "discourtesy complaint" had been referred to an internal charging committee for discipline, and that Rivieri had been assigned to Southeast District patrol since November 2008 (Kearney, 2009). Moses said, "The investigation's been completed. It's currently before the charging committee. What they'll do with it I don't know" (Augenstein, 2009). These statements appear to reflect the facts of Rivieri's duty assignments, but contradict official statements made December 12, 2008.

On September 27, a department spokesman stated that internal investigators sustained administrative charges against Rivieri of using excessive and unnecessary force and using discourtesy. The penalties for Rivieri could range from a reprimand to termination (Hermann, 2009e). At this time, Rivieri was offered the opportunity to plead guilty, accept the administrative punishment, and keep his job. A senior police source, speaking off the record due to personnel confidentiality rules, said that Commissioner Bealefeld had offered Rivieri a 90-day suspension, spread out to avoid his having to go a full three months without pay, and an agreement to attend anger management classes (Hermann, 2010f). Rivieri declined, and insisted on a trial board.

On July 20, 2010, Rivieri was cleared by a trial board chaired by Maj. Terrence P. McLarney, the commander of the homicide unit, who had been on the force 33 years. The three police officers comprising the board had held a hearing the preceding week. They found Rivieri not guilty of using excessive and unnecessary force and uttering a discourtesy, but guilty of failing to issue a citizen contact receipt and to file a report. The board recommended six days' suspension without pay and six days' loss of leave (Hermann, 2011).

Commissioner Bealefeld could have accepted the board's decision, or chosen a more severe punishment. He had the authority to escalate punishment within the parameters of a matrix; the offense of failing to write a report offers maximum discretion, including termination. Bealefeld reportedly was reviewing the video personally and had 30 days to decide what action to take (Janis, 2010). The board's ruling overturned the IID investigation, which had concluded that Rivieri had exceeded his authority in the incident (Hermann, 2011).

On August 25, 2010, at 9:30 a.m., Commissioner Bealefeld told Rivieri that he had brought discredit on the Police Department and the city, and that he was being terminated. Guglielmi confirmed the personnel action (Hermann, 2010 August 25). Police union officials later reported that Rivieri was called to headquarters from patrol, and then fired after a two-minute hearing in the commissioner's office. Rivieri was required to surrender his gun. He reportedly had to call a friend for a ride because he was not allowed to drive back in the department's squad car. Guglielmi later elaborated, "This administration has made it abundantly clear that we are going to hold people accountable. The people of Baltimore deserve that, and more. That's how we take internal policing. Very seriously" (Hermann, 2010d).

The variable of the **final outcome regarding the officer involved** was that Rivieri was terminated by the police commissioner (Hermann, 2010b). One comment posted to the *Sun*'s Crime Beat blog on August 27, apparently by someone with access to internal police information, neatly summarized the issue: "Rivieri had the chance to save his job but turned it down and requested a trial board hearing. The FOP is pissed because they can control the trial board but can't control the commish" (Hermann, 2010d).

On February 28, Circuit Court Judge Sylvester Cox ruled that Bealefeld did not act in an arbitrary or capricious manner. The judge noted that the trial board's recommendations are not binding, and that Bealefeld acted within his authority in firing Rivieri. "The court is not here to second-guess the police commissioner. The commissioner acted well within his discretion. This court is not going to disrupt his position" (Hermann, 2011).

For the variable of **final outcome of the law enforcement agency regarding policy changes, if any** include: changing the leadership and increasing the staffing of the Inner Harbor patrol unit, mandating additional training for all members of that unit (Linskey, 2008), ceasing the release of officer names when they are accused of shooting or assaulting civilians (Hermann, 2009a), and updating the civilian complaint website page (Baltimore City Police, 2008).

On October 27, 2010, the website for the Baltimore Police Department had a new page with a five-minute video about the Inner Harbor Bike Unit. The lieutenant in command of the unit stated that they needed to be "ambassadors of the city" because of the international tourism in the Inner Harbor park. He also said that bike unit officers had to be in shape and had to be able to speak well. "This is not a country club down here at the Inner Harbor, you do have to work hard" (Baltimore City Police, 2010).

The variable of the **affiliation of the civilian subject** appears to be entirely civilian; like the videographer, the subject is a high school student and skateboarder.

The variable of **the initial official response to handling of the civilian subject** was complex. On July 1, 2007, Rivieri was patrolling the Inner Harbor area. Skateboarding was prohibited in the area, and there were multiple signs posted (Cobus, 2008). Skateboarding was a citable offense, meaning that police had discretion to write a ticket but not to arrest an individual for skateboarding (Hermann, 2010b). Rivieri had been with the department since his Entrance On Duty (EOD) date of August 28, 1991, sixteen years (Skinner, 2008).

All accounts agree that prior to the recording, Rivieri had warned the teens to stop skateboarding, and Bush stated in a television interview that he was "100 percent wrong" (Jones & Sentementes, 2008) for skateboarding there. Rivieri then drove about 30 yards in his electric cart. He checked his rearview mirror and saw that at least one teen had ignored his warning (Hermann, 2009e). Bush later stated, "We were just skating there. The officer was rolling by and I was listening to my iPod and didn't hear him" (Good Morning America, 2008). Santo told the same story in the YouTube video comments:

> he told us to leave and we did. eric didn't hear it so he keap doin it. we told him to get off and he did. we were walkin away but know the cop backs up for no reason knowing were leaving he even sees us but no he gets out and the rest is in the video. (Santo, 2008)

It is apparent from the beginning of the video that Rivieri is already angry when he approaches the boys for the second time. Rivieri appears to interpret Bush's initial responses as disrespectful, and twenty seconds into the incident it is clear that Rivieri has lost control of his temper (Santo, 2008, 00:21:12). The next three minutes are mostly Rivieri shouting at Bush, including profanity and several threats. The ensuing BPD IID investigation characterized Rivieri's actions as using excessive and unnecessary force and uttering a discourtesy (Hermann, 2010f). 43 seconds into the video, Rivieri grabs Bush's skateboard, grabs Bush around the head, and pushes him to the ground, then pushes him to the ground again. Rivieri then turns his back on the group of teens, walks back to his cart, and throws the skateboard into the back (Santo, 2008, 01:11:27). On returning to the teens, Rivieri resumes his shouting, including a strong negative response to Bush's use of the word "dude" (Santo, 2008, 02:42:21). At three and a half minutes, Rivieri apparently notices Santo holding the video camera, and immediately threatens, "Y'got that camera on? If I find myself on Y-" (Santo, 2008, 03:35:19).

Rivieri reportedly allowed Bush to call his mother, Peggy Miller, after the camera stopped recording. Rivieri told Miller that Bush was being disrespectful. Rivieri returned the skateboard and left a few minutes after the end of the video. Bush later recalled, "I was pretty scared. I was thinking he was going to do something else, punch me in my face" (Jones & Sentementes, 2008).

Rivieri's court filings state that he "shook hands with the young man" and "issued him a Citizen Contact Form" (Hermann, 2009e). While the handshake might have occurred, IID and the police review board both determined that Rivieri had not in fact issued Bush a Citizen Contact receipt. Department regulations require that, for incidents such as this one, police issue a Citizen Contact receipt to the citizen, and file a report. Rivieri did neither, leaving no official paper trail for the incident; only Santo's videotape remains as evidence.

The variable of the **sequence of official actions regarding the civilian subject** was that once the department became aware of the video, police told reporters that they did not know the identities of the boys involved, and that Rivieri had been suspended pending investigation (Linskey & Sentementes, 2008; Skinner, 2008). Bush participated in several interviews with IID detectives (Jones & Sentementes, 2008). It is not clear whether he was ever asked to testify at internal hearings. His civil suit was dismissed before trial began, so he was never asked to testify.

The variable of the **compensation (if any) of the subject** is negative. There are no public documents that indicate Bush was ever compensated in any way; on the contrary, the ruling dismissing his civil suit assigned costs to plaintiffs, he and his mother (*Miller v. Rivieri*, 2008).

The variable of the **final outcome regarding the civilian subject** was the dismissal of his civil lawsuit (Hermann, 2009c).

The variable of **what third parties (if any) involved themselves in the legal case** is that no third party involved itself on behalf of Bush, and the Fraternal Order of Police, Lodge #3, the local police union, is the only third party that involved itself on Rivieri's behalf (WBAL, 2010a).

The variable of ***cui bono?*, or who benefits?**, in this case has several answers. Rivieri does not benefit, having lost his job, his pension, and his reputation. Eric Bush does not appear to have profited financially, but has seemed to enjoy some of his notoriety, as witness the DUDE tattoos on his wrists (WBAL, 2010c). Tony Santo has evidently appreciated the notoriety of his video, as attested by his comments on his YouTube channel (Santo, 2010). The BPD has, on balance, probably benefitted by earning a slightly better reputation for transparency at the highest levels. The citizens of Baltimore will have benefitted by a more accountable police department, if the BPD actually follows through with its reformed policies on investigating civilian complaints. Skateboarders in Baltimore may benefit on balance, as there is now more public pressure for the construction of skate parks, and this may compensate for any increase in anti-skating enforcement.

The variable of **how many times the video was viewed online** was over 400,000 times within a few days of being posted, and as of February 1, 2012 over five and a half million times (Santo, 2008). Several of the other videos related to Rivieri's misconduct had been viewed over one million times each (Exzellent, 2008; TechBalt, 2008; JaviErick, 2008), for ten million total.

The variable of **how much press coverage the event received** included all major television networks, multiple radio stations, AP and UPI newswires, and national, regional, and local newspapers, with 49 articles on ProQuest Newsstand, and 20 on LexisNexis Academic.

The variable of **how much Internet discussion was linked to this event** is that there were over 78,000 comments on camb0i's YouTube video alone (Santo, 2008), over 800 blogs, and 1,500 forum threads (Google +Rivieri +Bush).

CASE STUDY II: ISMAIL, LONG, POGAN, 2008-07-25

On July 25, 2008, on 7th Avenue near 46th Street in Manhattan, civilian Asam Ismail videorecorded NYPD Officer Patrick Pogan pushing civilian cyclist Christopher Long off his bicycle and onto the street. This case is included for direct replication (similar results for expected reasons), per Yin (2003, pp. 5, 110); the phenomenon is a police-civilian interaction in American public space during which police misconduct may have occurred, and of which user-generated video was posted online.

Research Question and Hypotheses

For Research Question 1, what is the outcome of user-generated online video on police-civilian interactions in American public space?, the outcomes of this case include: assault charges against Long were dropped (Eligon & Moynihan, 2008); the suspension, resignation and felony conviction of Pogan (Weiss et al., 2008; *The New York Times*, 2009; Eligon, 2010); a \$65,000 plus \$25,000 for attorney's fees settlement for Long's civil rights suit (*Long v. The City of New York et al.*, 2010); no admission of guilt from the NYPD or city (*Long v. The City of New York et al.*, 2010); a payment of \$310 to Ismail for the video (Peltz & Long, 2010), minus the inconvenience of returning to New York to testify (Khan, 2010); a contributing factor to another \$965,000 settlement to cyclists in a class-action suit (NY1 News, 2010).

For this case, Hypothesis 1, that user-generated online video has the potential to improve accountability in police-civilian interactions in American public space, appears to be supported. Police falsified a charging document that was almost completely at odds with events as they appear in the video. The discrepancies between that document and the video could not be reconciled by the officer's defense. Thus, an NYPD officer was convicted of a felony (Eligon, 2010), which bars him from ever re-applying to the department (Grace, 2010c), and the city paid significant compensation both to the civilian in this case (*Long v. The City of New York et al.*, 2010) and to other cyclists in a related class-action suit (NY1 News, 2010).

The value of Hypothesis 2, that user-generated online video is significantly different from professional video journalism in its effects on accountability in police-civilian interactions in American public space, in this case does not appear to be evaluable without conjecture.

Hypothesis 3, that there are strong motivations for police to continue to attempt to restrict civilians' First Amendment rights to photograph police in public spaces, appears to be supported; a police career was ended by the evidence of user-generated online video.

Documents

The video was shot during a mass cyclist protest; the cyclist organization had many cameras of their own, but a tourist from Florida got the best recording. The organization trimmed the beginning and end of the video, added titles, posted it to YouTube (downsouthvids, 2008), and alerted the media (Weiss et al., 2008). A second video showing other officers involved was later made public through being introduced as evidence in court (*New York Post*, 2010).

The misdemeanor complaint signed by Pogan, accusing Long of assaulting him with a bicycle, is the piece of evidence sealing Pogan's felony conviction for falsifying business records and offering a false instrument for filing (Eligon, 2010; *New York v. Long*, 2008). This bars him from ever re-applying to the NYPD (Grace, 2010c). This document was first made public as a PDF file on the Internet, and remains available from multiple locations (*New York v. Long*, 2008).

Significant court documents include Long's misdemeanor charging documents, including the dismissal of all charges (*New York v. Long*, 2008); Long's civil rights suit complaint, the city's answer, and the final settlement documents (*Long v. The City of New York et al.*, all dates); trial testimony, particularly that of Sgt. Eric Perez, who advised Pogan (Italiano, 2010a); and Pogan's felony conviction ana sentencing (Italiano, 2010b; Yaniv, 2010). Because the case went to trial, many more details became public record, including Pogan's and Long's backgrounds. This increased the number of details available, but multiplied the number of pro-forma court documents to be sifted for very few data pertinent to the variables of interest.

Pogan had only been on duty eleven days (Eligon, 2010; Weiss et al., 2008), so it is not surprising that no other civilians came forward with similar complaints, and that there were no documents about his prior incidents. However, additional video from the cyclist organization revealed that senior NYPD personnel on the scene with Pogan had performed similar actions at other mass rides (Glass Bead Collective, 2007).

The New York media have covered this case thoroughly, including many direct quotes from courtroom testimony. The case data exceed 80 files, including 23 media clips from YouTube, broadcast TV, radio, and NYPD official releases. The court documents are numerous, but relatively few are crucial. There were 53 articles from LexisNexis

Academic News alone, and the search +Pogan +Long +video retrieves more than 70 Google News articles (archives included). There have been two law review articles that cite the case.

Relevant Variables of Interest

It is evident from Ismail's video that the police never looked in the direction of the camera or otherwise indicated that they were aware of it or of the actions of the videographer, so the variables of **whether the police were aware of the camera, when the police became aware of the camera, whether there was any marked change in police behavior once they were aware of the camera, whether police made any attempt to prevent the recording, whether police made any attempt to acquire, confiscate, or destroy the video, whether the police gave any unlawful instruction to the videographer regarding the video**, and **whether police detained, cited, or arrested the videographer** are all negative (downsouthvids, 2008).

The variable of **when the video was available via the Internet** is July 27, 2008, two days after the incident, when the video was uploaded to the YouTube channel of downsouthvids, where it has remained available (downsouthvids, 2008).

The variable of **when the video was available via broadcast news media** is no later than July 28, 2008, at 7:24 PM, when it was broadcast on NY1 (NY1 News, 2008).

The variable of when images from the video were available via print news media is July 28, 2008 (Weiss et al., 2008).

The variable of **whether more than one camera captured the event** is positive; there were other civilians operating cameras visible in the original video, and a second video of the incident was later made public and posted online (downsouthvids, 2008; *New York Post*, 2010).

The variable of **whether official CCTV, dashboard, or other video captured the event** is apparently negative. Despite the fact that Manhattan has a very high density of official and commercial CCTV cameras that are linked through a police-controlled system (NYCLU, 2006), officials never produced or admitted to the existence of official video of the event. Thus, the variables of **whether police initially admitted to the existence or possession of video of the event, when did police admit to possession of video of the event, whether police**

released official video of the event, and **whether official video of the event was available via the Internet** are all not applicable.

The variable of **who was credited as the source for each medium of release of the video** is for most instances either YouTube, or anonymous tourist (Weiss et al., 2008; NY1 News, 2008; Eligon, 2008); the videographer was apparently not credited by name in any medium of release prior to Pogan's criminal trial (Peltz & Long, 2010; Khan, 2010). The video is credited on the original YouTube posting to "an anonymous tourist and members of the Time's Up! Environmental Group, Glass Bead Collective, and I-Witness Video" (downsouthvids, 2008).

The variable of **whether there was any effort to restrict, remove or prosecute the release of the video** is mostly negative, although YouTube imposed age restrictions for viewing the original video, based on the company's community guidelines (downsouthvids, 2008; YouTube, 2011).

The variable of the initial official response to handling of the videographer is none.

The variable of the **final outcome regarding the videographer** is slightly positive; although he had the inconvenience of testifying at Pogan's trial, he was paid for his video (Peltz & Long, 2010; Khan, 2010).

The variable of the **sequence of official actions regarding the videographer** is that he was called to testify at Pogan's criminal trial (Khan, 2010).

The variable of the **compensation of the videographer** is positive; demonstration organizers asked bystanders for video of this incident, and paid Ismail $310 for his tape (Peltz & Long, 2010).

The variable of the **affiliation of the videographer** appears to be civilian. He responded to demonstration organizers' requests for bystander videos (downsouthvids, 2008; Peltz & Long, 2010). However, in a television interview he stated that his "three kids are in a police department right now" (Khan, 2010, 00:20).

The variable of **whether police misconduct was recorded** is positive; although Pogan was later acquitted of most of the charges, the video justified an investigation. The former offensive lineman made a rather spectacular hit on the cyclist, which brought up the question of appropriate use of force (downsouthvids, 2008; *New York v. Long*,

2008; *Long v. The City of New York et al.*, 2010; Weiss et al., 2008; Eligon & Moynihan, 2008; Eligon, 2010).

The variable of the **initial response of the law enforcement agency regarding the officers involved** was to put Pogan on desk duty as of July 28, pending investigation (Weiss et al., 2008; NY1 News, 2008).

The variable of the **sequence of official actions regarding the officers involved** was complex. NYPD Internal Affairs Division investigated the incident, following which the Manhattan district attorney's office indicted Pogan on December 16 for assault and filing false paperwork (*Long v. The City of New York et al.*, 2009a; Eligon & Moynihan, 2008). Pogan resigned from the NYPD in early February 2009, just before he was to be fired. Throughout this period, city and police officials publicly expressed concern at the video, but also expressed regret that a third-generation officer was in trouble (*Long v. The City of New York et al.*, 2009a; Eligon, 2010; *The New York Times*, 2009).

Pogan's criminal trial in Manhattan Supreme Court began April 19, 2010 (Peltz & Long, 2010). The videographer testified the first day (Khan, 2010). In testimony April 23, Sgt. Eric Perez, who prepared the charging paperwork that Pogan signed, stated that Pogan "just said that the bicycle had struck him and caused him to fall down" (Italiano, 2010a). Perez also stated that in instructing Pogan and seven other probationary officers, "I told them to let the detail enforce the laws. We wanted them there as backup, not to actually engage and stop" (Grace, 2010a). The same day, Assistant District Attorney Laura Millendorf testified that Pogan informed her that Long had ignored his commands to stop and "turned his bike around" toward Pogan. Millendorf stated, "I thought the act of charging head on into a police officer was egregious and should be the top count" (Italiano, 2010a).

On April 29, the jury found Pogan not guilty of assault or of harassment, but guilty of two felonies: falsifying business records and offering a false instrument for filing (Eligon, 2010; Grace, 2010c). Prior to sentencing, Assistant DA Ryan Connors, who had cross-examined Pogan, called him a perjurer and asked the court for jail time: "He told a blameless story of fantasy where he did nothing wrong and everyone around him had distorted the truth" (Italiano, 2010b).

The variable of the **final outcome regarding the officers involved** was that Pogan received no jail time, but is barred by his felony conviction from both the NYPD and the NYFD (Grace, 2010c; Italiano, 2010b). The sentence was lighter than the defense had requested; they expected community service. At sentencing, Manhattan Supreme Court Justice Maxwell Wiley said, "The defendant doesn't need any further supervision by the court and the verdict is conditional discharge, period" (Yaniv, 2010).

For the variable of **final outcome of the law enforcement agency regarding policy changes, if any**, there has been no admission of guilt; no policy changes have been evident (*Long v. The City of New York et al.*, 2009c, 2010).

The variable of the **affiliation of the civilian subject** is that he was a member of an activist group (Critical Mass) known for demonstrations that some have characterized as provoking the police. The incident occurred during one of those demonstrations (Weiss et al., 2008; Eligon & Moynihan, 2008; Peltz & Long, 2010; Grace, 2010b).

The variable of **the initial official response to handling of the civilian subject** is complex. Around 9:30 PM on July 25, 2008, during a cyclist demonstration on 7th Avenue near 46th Street in Manhattan, rookie NYPD Officer Patrick Pogan knocked cyclist Christopher Long off his bicycle and onto the street (downsouthvids, 2008; *Long v. The City of New York et al.*, 2009a). On the video, it is evident that Long was attempting to avoid Pogan, that Pogan took several steps to intercept Long, and that Pogan remained on his feet after hitting Long. It is not evident that Pogan issued any instructions to Long prior to striking him (downsouthvids, 2008; Weiss et al., 2008). Pogan and other officers then arrested Long (Weiss et al., 2008; Eligon, 2010; *New York v. Long*, 2008). Pogan thereafter signed documents charging Long with attempted assault in the third degree, resisting arrest, and disorderly conduct, and claiming that Long ran down Pogan with his bicycle (Weiss et al., 2008; Eligon, 2010; *New York v. Long*, 2008; *Long v. The City of New York et al.*, 2009a). The relevant passage of that document states:

> …the defendant steered the defendant's bicycle in the direction of deponent and drove defendant's bicycle directly into deponent's body, causing deponent to fall to the ground and causing deponent to suffer lacerations on deponent's forearms. (*New York v. Long*, 2008)

Police fingerprinted Long and kept him in a cell in the Midtown South precinct house for several hours, then transferred him to Manhattan Central Booking, where he spent the next 24 hours detained with general population arrestees. He was arraigned on the three charges at 2 AM Sunday, July 27, and was released on his own recognizance (Weiss et al., 2008; *Long v. The City of New York et al.*, 2009a, pp. 4-7).

The variable of the **sequence of official actions regarding the civilian subject** is complex. On September 5, at Long's first court appearance, the Manhattan district attorney's office dropped all charges against him, citing a lack of evidence (*New York v. Long*, 2008; *Long v. The City of New York et al.*, 2009a). On July 7, 2009, Long filed a civil rights suit against Pogan and the city, asking for $1.5 million (*Long v. The City of New York et al.*, 2009a). In October, the city settled with Long before trial, with the defendants making no admissions or apologies. The city also delayed full payment until December 2010, after more than a year of motions over Long's attorneys' fees (*Long v. The City of New York et al.*, 2009c, 2010).

The variable of the **compensation (if any) of the subject** is $65,001 plus $25,000 in attorney's fees. Although Long settled in October 2009, the city delayed full payment until December 2010 (*Long v. The City of New York et al.*, 2009c, 2010).

The variable of the **final outcome regarding the civilian subject** was a settlement for $65,001 plus $25,000 in attorney's fees, with no admissions or apology from the city or the police (*Long v. The City of New York et al.*, 2009c, 2010).

The variable of **what third parties (if any) involved themselves in the legal case** is Critical Mass, a cyclist rights group (*New York Post*, 2010; downsouthvids, 2008).

The variable of ***cui bono?*, or who benefits?**, in this case is evidently the civilian subject and other cyclists in New York City (NY1 News, 2010).

The variable of **how many times the video was viewed online** is over 2.8 million as of February 2, 2012; it had been viewed 1.2 million times as of July 31, its fourth day online. It is worth noting that the video has been age-restricted based on YouTube community guidelines (downsouthvids, 2008; Wasserman, 2009, p. 605, n.24; YouTube, 2011).

The variable of **how much press coverage the event received** was at least in the upper tens of news reports. The New York City media covered the incident most thoroughly, but regional, national, and international news outlets carried the story as well. There were 53 articles from LexisNexis Academic News alone, and the search +Pogan +Long +video retrieves more than 70 Google News articles (archives included). Nearly every news story throughout the active life of this case included still frames, a copy of the video, or a link to the video on YouTube (downsouthvids, 2008; Weiss et al., 2008; NY1 News, 2008; Eligon & Moynihan, 2008; Eligon, 2010; Peltz & Long, 2010; *New York Post*, 2010).

The variable of **how much Internet discussion was linked to this event** is over 10,000 comments on the original YouTube video alone (downsouthvids, 2008). A Google search for +Pogan +Long +video retrieves over 1700 blogs, over 1500 discussions, and more than 240 videos.

CASE STUDY III: HURLBUT, SMOKER, TROLLEY GUARDS, 2009-09-05

On September 5, 2009, civilian Rob Hurlbut videorecorded six San Diego trolley guards arresting a smoker at the 12th and Imperial transit station, a public space. This case is included for direct replication (similar results for expected reasons), per Yin (2003, pp. 5, 110). The phenomenon is a police-civilian interaction in American public space during which police misconduct may have occurred, which was videorecorded by a civilian, and the videorecording was distributed online.

It is particularly important to examine this case because it exemplifies attempted prior restraint on First Amendment protected speech by police. This is precisely the point raised in the first justification for the present research: The stated problems have significant implications for the continued exercise of First Amendment rights in photographing public space, both for autonomous citizens and for professional journalists.

The scope of the present research explicitly includes police. To reiterate, historically and recently, the state has extended police powers to a broad range of persons. Thus, the term 'police' as used in this research includes any law enforcement officer, private security guard, or other person granted police powers to act for a government agency while interacting with civilians in American public space. As previously noted, present estimates are that each public police officer has approximately three counterparts in private security (Goldstein, 2007, para. 7). There has also been a trend to the privatization of policing, even of public spaces such as parks and transit stations (Mitchell, 2003, p. 1). In this case, the government agency is the MTS:

> Metropolitan Transit System (MTS) is a California public agency and is comprised of the San Diego Transit Corp. and San Diego Trolley, Inc., nonprofit public benefit corporations, in cooperation with Chula Vista Transit and National City Transit. MTS is the taxicab administrator for eight cities. MTS is the owner of the San Diego and Arizona Eastern Railway Company. (Metropolitan Transit System, n.d.)

The same document lists MTS member agencies as "The cities of Chula Vista, Coronado, El Cajon, Imperial Beach, La Mesa, Lemon Grove, National City, Poway, San Diego, Santee, and the County of San Diego" and states, "MTS officers in uniform and in plainclothes routinely patrol trains, buses, stations, and parking lots" (Metropolitan Transit System, n.d.). At the time of the incident, MTS outsourced its security services to a privately held firm, Heritage Security Services, which also does business as Transit Systems Security (San Diego Trolley, Inc., 2006, p. 1).

MTS transit stations, and specifically the 12th and Imperial transit station where this incident occurred, are American public space within the scope of this research. Ken Moller, President of Heritage Security Services, stated "It's a public place…" (Hurlbut, 2009c, 2009d). Thus, Heritage Security Services employees are granted police powers to act for a government agency while interacting with civilians in American public space, and are therefore police within the scope of the present research.

Research Question and Hypotheses

For Research Question 1, what is the outcome of user-generated online video on police-civilian interactions in American public space?, the outcomes of this case include: a public statement of error and a clarification of policy by the head of a private security company (Hurlbut, 2009c); and refusal of that company (and the public agency it serves) to release pertinent documents (Snyder, 2010).

For this case, Hypothesis 1, that user-generated online video has the potential to improve accountability in police-civilian interactions in American public space, appears to be supported. The publication of the video prompted the company's apology and clarification (Hurlbut, 2009a, 2009c). Without the user-generated online video, there would have been no incentive for the security company to issue corrective instruction to its employees.

The value of Hypothesis 2, that user-generated online video is significantly different from professional video journalism in its effects on accountability in police-civilian interactions in American public space, in this case does not appear to be evaluable without conjecture.

Hypothesis 3, that there are strong motivations for police to continue to attempt to restrict civilians' First Amendment rights to photograph police in public spaces, does not appear to be evaluable without conjecture. There are no documents recording definite sanctions for the trolley guards.

Documents

Of the 24 documents for this case, only two – the newspaper story (Snyder, 2010) and the broadcast television report (Hurlbut, 2009c) – are from traditional mass media. The remaining documents are from websites, blogs, and online forums. There are no court documents that can be specifically linked to this incident at this time because officials have refused to release any names.

Relevant Variables of Interest

The variable of **whether the police were aware of the camera** is positive. In the video, the female guard looks directly into the camera and visibly changes posture, direction, and speed of movement in the camera's direction (Hurlbut, 2009a).

The variable of **when the police became aware of the camera** is 00:15, when the female guard looks directly into the camera and visibly changes posture, direction, and speed of movement in the camera's direction (Hurlbut, 2009a).

The variable of **whether there was any marked change in police behavior once they were aware of the camera** is positive. The female guard visibly changes posture, direction, and speed of movement in the camera's direction as soon as she becomes award of it, and also immediately begins issuing unlawful orders to the videographer (Hurlbut, 2009a).

The variable of **whether police made any attempt to prevent the recording** is positive. In the video, two guards repeatedly order Hurlbut to stop taking pictures, and repeatedly interpose themselves between Hurlbut and the guards restraining the smoker (Hurlbut, 2009a).

The variable of **whether police made any attempt to acquire, confiscate, or destroy the video** is negative; there is no evidence that any of the guards made any attempt to acquire, confiscate, or destroy the video.

The variable of **whether the police gave any unlawful instruction to the videographer regarding the video** is positive. In the video, two guards repeatedly order the videographer to stop taking pictures, which orders are unlawful as the location is a public space (Hurlbut, 2009a).

The variable of **whether police detained, cited, or arrested the videographer** is negative. There is no evidence that the guards detained, cited, or arrested the videographer. However, the videographer stated that the guards threatened him after the camera was turned off (Hurlbut, 2009b).

The variable of **when the video was available via the Internet** is September 7, 2009, when the videographer uploaded the video to his YouTube channel (Hurlbut, 2009a, 2009b).

The variable of **when the video was available via broadcast news media** is Friday, September 18, at 11 p.m. on San Diego NBC affiliate 7/39, including a brief interview with Hurlbut and the text of a response from Ken Moller, President of Heritage Security Services (Hurlbut, 2009c, 2009d).

The variable of **when images from the video were available via print news media** is apparently never, based on the document search and acquisition procedures. There is no evidence that images from the video have appeared in any print publication, including the *San Diego Reader* article (Snyder, 2010).

The variable of **whether more than one camera captured the event** is positive. At approximately 01:00 in the video, another bystander can be seen tracking the guards with what appears to be a cell phone camera. However, any video she may have recorded has apparently not been made public (Hurlbut, 2009a).

The variable of **whether official CCTV, dashboard, or other video captured the event** is not applicable. It is possible that a CCTV, MTS security dashboard, or other official camera captured the event (Clock, 2009), but any such recording has never been released to the public, and MTS has never admitted the existence or possession of video of the event (Snyder, 2010). Therefore, the variables of **whether police initially admitted to the existence or possession of video of the event, when did police admit to possession of video of the event**, **whether police released official video of the event**, and **whether official video of the event was available via the Internet** are apparently all not applicable.

The variable of **who was credited as the source for each medium of release of the video** is that Hurlbut was credited as the source of the video for the broadcast, and by his online alias of 'theworldisraw' on his YouTube channel (Hurlbut, 2009a, 2009b, 2009c, 2009d).

The variable of **whether there was any effort to restrict, remove or prosecute the release of the video** appears to be negative. The videographer has not reported any such attempts (Hurlbut, 2009b, 2009d; Snyder, 2010).

The variable of the **initial official response to handling of the videographer** was that Ken Moller, President of Heritage Security Services, replied to the television station's questions prior to the September 18 broadcast in what amounted to an apology to the videographer:

> We have no right to tell people they can't shoot (video) down there. My officers were wrong in telling him that. And I put the word out as soon as I saw the video. It's a public place and people can certainly shoot video down there if they want to. (Hurlbut, 2009c, 2009d)

The variable of the **sequence of official actions regarding the videographer** is complex, but is mostly documented by the video. At 8:20 p.m., trolley guards detained a man for smoking. Hurlbut witnessed the incident and described it in a comment posted to his website on September 19 (Hurlbut, 2009b), and provided additional details in a comment to a report of the story on Carlos Miller's *Pixiq* (formerly *Photography Is Not A Crime*) website. The smoker was reportedly exiting the station toward the smoking area, was interrupted by a trolley, and did not immediately extinguish his cigarette when the trolley guards instructed him to do so. The guards reportedly chose to forcibly detain the smoker rather than issue a citation per the city ordinance (San Diego Mass Transit System, 2007). "When the two original guards started to take him down, and were quickly joined [by] a couple other guards, and that's when I started filming" (Miller, 2010a).

The video documents that the female guard says to the videographer, "Hey, hey get away with the camera. Hey. You're not allowed to take pictures" (Hurlbut, 2009a, 00:15). The female guard and the largest guard then interpose themselves between the videographer and the guards subduing the smoker. After the guards remove the smoker to an MTS vehicle, the female guard looks back toward the videographer (02:00) and immediately moves to talk with other guards. The largest guard then moves to confront the videographer, asks to see his ticket, then says, "All right, we've asked you not to take pictures, so, no taking pictures" (Hurlbut, 2009a, 02:28) In the ensuing conversation, the videographer repeatedly asks if taking pictures or video is against the law. The guard simply repeats, ten times, that the guards don't want any pictures taken, and that the videographer is to stop (Hurlbut, 2009a, 03:00). It is worth noting that the videographer was using a Digital Single-Lens Reflex (DSLR) camera that had a video mode (Hurlbut, 2009b). From their repeated use of the phrase 'no taking pictures' it is evident that the guards perceived it as only a still camera, and did not realize that he was recording them on video.

The videographer later stated that he knew he was legally in the right to keep recording, but that he felt he already had "some great footage" and was afraid the guards would arrest him and confiscate or destroy it (Snyder, 2010, paras. 18-19). The videographer reported that a few minutes later, another guard asked to see the videographer's pass, and asked where he was going. When the videographer replied that he was headed to La Mesa on the orange line, the guard told him to wait near where that trolley would leave. As the videographer moved to comply, that guard reportedly instructed another guard to ensure that the videographer got on the next trolley. "I interpreted the exchange to mean that something would happen if I wasn't on the next trolley. I took his meaning to be that they wanted me to clear out of the station" (Snyder, 2010, paras. 20-21). The videographer said that he left on the 8:34 trolley.

After the video was available on YouTube and had been picked up for a story on San Diego NBC affiliate 7/39, the president of the security company responded to station's request for comment with the statement that the guards had been wrong (Hurlbut, 2009c, 2009d).

Hurlbut followed up by emailing MTS. He received an automated reply from the email account of Belinda Fragger (belinda.fragger@sdmts.com): "Thank you for bringing this matter to our attention. Your email has been forwarded to MTS Trolley for handling. MTS case #41411." As of the time of publication of Snyder's article, Hurlbut had not heard back from anyone at MTS, or from Fragger (Snyder, 2010, paras. 29-30).

Thereafter, both the security company and the Metropolitan Transit System refused requests for the incident reports from that evening. Tiffany Lorenzen, general counsel for MTS, justified the refusal based on the California Public Records Act, which exempts from disclosure documents that are: (1) records pertaining to current litigation to which the public agency is a party; or (2) records of complaints to, or investigations conducted by the Office of the Attorney General and the Department of Justice, and any state, or local police agency. Lorenzen refused to be more specific as to the reason for the refusal (Snyder, 2010, paras. 32-33).

The variable of the **final outcome regarding the videographer** is that the security company president apologized, but that the videographer was unable to elicit any further information about the incident (Hurlbut, 2009c, 2009d; Snyder, 2010; Miller, 2010a, 2010c).

The variable of the **compensation of the videographer** is: compensation of the videographer was not evident, based on the document search and acquisition procedures. The number of views of the video on YouTube may be in the range where the videographer can earn revenue from advertising, but he has apparently not monetized his channel at this time (Hurlbut, 2009b).

The variable of the **affiliation of the videographer** is that he is a civilian freelance photographer who was not on assignment at the time (Hurlbut, 2009b, 2009c, 2009d; Snyder, 2010).

The variable of **police misconduct** is positive; the trolley guards issued unlawful instructions, which was admitted to and apologized for by their employer (Hurlbut, 2009c, 2009d).

The variable of the **initial response of the law enforcement agency regarding the officers involved** was that Ken Moller, President of Heritage Security Services, replied to the television station's questions prior to the September 18 broadcast:

> We have no right to tell people they can't shoot (video) down there. My officers were wrong in telling him that. And I put the word out as soon as I saw the video. It's a public place and people can certainly shoot video down there if they want to. (Hurlbut, 2009c, 2009d)

Hurlbut posted a copy of the broadcast video segment to his YouTube channel as "San Diego Trolley Police Were Wrong" on September 21, then linked the video to his photography website, *The World is Raw* (Hurlbut, 2009c, 2009d).

Attempts by the videographer and by journalists to gather information subsequent to the incident were frustrated when both the security company and the Metropolitan Transit System refused requests for the incident reports from that evening (Miller, 2010c; Snyder, 2010). Thus, the variables of interest including the **sequence of official actions regarding the officers involved**, the **final outcome regarding the officers involved**, and the **final outcome of the law enforcement agency regarding policy changes, if any** are unknown because the information has been withheld.

The variable of **the initial official response to handling of the civilian subject** is that at 8:20 p.m. on Saturday, September 5, 2009, six trolley guards forcibly detained a man for smoking (a $75 citable infraction of an MTS ordinance) at the 12th and Imperial transit station (Hurlbut, 2009a; San Diego Mass Transit System, 2007; Miller, 2010a). Again, due to the complete denial of information from both MTS and Heritage, the variables of interest including the **affiliation of the civilian subject**, the **sequence of official actions regarding the civilian subject**, the **compensation (if any) of the subject**, and the

final outcome regarding the civilian subject are unknown because the information has been withheld. The variable of **what third parties (if any) involved themselves in the legal case** is also unknown, as it has not been possible to identify any legal case using the document search and acquisition procedures.

The variable of ***cui bono?*, or who benefits?**, for this case has several possible answers. The mass transit users of San Diego may benefit if the security company president's apology is interpreted as evidence of a more accountable security service. Photographers and videographers may benefit if trolley guards respect the public statement that using cameras is legal in MTS transit stations. However, the lack of transparency persisting about the original incident argues that Heritage Security Services has benefitted by the rapid and thorough burying of the story (Snyder, 2010; Hurlbut, 2009c; Miller, 2010c).

The variable of **how many times the video was viewed online** is 18,879 as of December 20, 2011 (Hurlbut, 2009b).

The variable of **how much press coverage the event received** is in the low single digits, and included the original television broadcast, and one newspaper article. The *San Diego Reader* revisited the story in detail in its February 24, 2010 issue (Snyder, 2010). Since that story, there has been no other media coverage. There has been no coverage of this case in the San Diego *Union-Tribune*, or the affiliated 4SD television station, both owned by Cox Communications. There has been no coverage of this case in the *North County Times - San Diego & Riverside Counties* (http://www.nctimes.com). There were no relevant stories on LexisNexis Academic, ProQuest Newsstand, or Access Newspaper Archive. There were no response videos or other incident-related videos on YouTube. There are no entries in Justia.com.

The variable of **how much Internet discussion was linked to this event** is in the tens of threads, which is low in comparison to most of the other cases. Internet discussion linked to the event has been concentrated in Hurlbut's YouTube channel, Carlos Miller's *Pixiq/Photography is Not a Crime* website, and the website of the *San Diego Reader* (Hurlbut, 2009b, 2009d; Snyder, 2010; Miller, 2010a, 2010b, 2010c), with links to those sources appearing in a range of blogs and forums.

CASE STUDY IV: VARGAS ET AL., GRANT, MEHSERLE, 2009-01-01

Shortly after 2 AM on January 1, 2009, on Platform 1 of the Bay Area Rapid Transit (BART) Fruitvale station in Oakland, CA, at least five civilian videographers recorded an incident in which BART Police Officer Johannes Mehserle shot Oscar J. Grant III. This case is included for direct replication (similar results for expected reasons), per Yin (2003, pp. 5, 110); the phenomenon is a police-civilian interaction in American public space during which police misconduct may have occurred, and of which user-generated video was posted online.

Research Question and Hypotheses

For Research Question 1, what is the outcome of user-generated online video on police-civilian interactions in American public space?, the outcomes of this case include: the officer's resignation, conviction, and jailing for involuntary manslaughter (*California v. Mehserle*, 2010, February 19, p. 5; Winston, 2011); the resignation, firing, and reinstatement of other BART police (Bulwa, 2010a; Bulwa, 2010d); numerous public protests and related damage in Oakland and San Francisco (*California v. Mehserle*, 2009b, pp. 14-17, 26); settlements of $1.5 million to Grant's five-year-old daughter (Bulwa, 2010c) and $1.3 million to Grant's mother (*Johnson et al. v. Bay Area Rapid Transit et al.*, 2011, p. 8; Bulwa, 2010c); civil suits by Grant's other relatives and friends (*Johnson et al. v. Bay Area Rapid Transit et al.*, all dates); a federal investigation into the incident (Gonzales, 2010); and an independent police auditor (BART, 2011).

For this case, Hypothesis 1, that user-generated online video has the potential to improve accountability in police-civilian interactions in American public space, is evidently positive, for holding Mehserle, the other police, and BART at least partially accountable for Grant's death.

The value of Hypothesis 2, that user-generated online video is significantly different from professional video journalism in its effects on accountability in police-civilian interactions in American public space, in this case does not appear to be evaluable without conjecture.

Hypothesis 3, that there are strong motivations for police to continue to attempt to restrict civilians' First Amendment rights to photograph police in public spaces, is positive, as the video evidence successfully countered the police 'blue wall' cultural practices of silence and perjury that were employed to protect Mehserle and other police from accountability for their actions.

Documents

There are presently more than 370 documents in hand, including 14 videos and more than 170 news wire reports, in addition to the court documents. Most of the important documents for this case were introduced at Mehserle's trial, and became part of the public record at that time. Activist websites such as Indymedia made many of the court documents, including preliminary hearing transcripts, available online. The intense media scrutiny made each new document release easier to identify. Six videos of the incident were introduced at Mehserle's trial: one from the BART platform security camera, and those recorded by five civilian witnesses. The synchronization and presentation of these videos in court was the subject of much legal wrangling; the judge finally allowed both sides to present their own videos. A single-screen synchronized version of the six videos has been available on YouTube since July 4, 2010. Almost all of the relevant BART internal documents were not publicly released until introduced in court. The court cases include multiple separate dockets: Mehserle's criminal trial, and several civil suits by Grant's relatives and friends; the civil suits were eventually combined. The CalCrim online system provided most of the pertinent court documents in this case, and the PACER system provided the remainder. The major civil suits are settled as of this date, but there is a remnant of final outcome data that may not be finalized for some time. A major challenge for this case is the sheer volume of data. Media activity continues; this is the most active of the cases studied. A second challenge is that a number of the incident participants and witnesses have chosen to speak through the alternative press. The differences in journalistic standards make it more difficult to discern accurate data, and differences in distribution have made access more challenging, as well.

Relevant Variables of Interest

The variable of **whether the police were aware of the cameras** is positive. The actions and statements of several of the police directly addressed one or more videographers (GioSifaTaufa, 2009, 07:25; *Los Angeles Times*, 2010; monkeyassj, 2009; ajajaj1, 2009; TheDirtyNews,

2009). Furthermore, Grant was able to use his cell phone camera to take a picture of Mehserle at close range shortly before he was shot, and Pirone told Grant, "You can't take pictures" (Grant, 2009; *California v. Mehserle*, 2009a, p. 1058), so it is unlikely that Mehserle was unaware of the cameras.

The variable of **when the police became aware of the camera** is that the videos appear to show that the police were aware of cameras throughout the incident (*Los Angeles Times*, 2010; monkeyassj, 2009; ajajaj1, 2009; TheDirtyNews, 2009).

The variable of **whether there was any marked change in police behavior once they were aware of the camera** is apparently negative, but only because Pirone's use of force and profanity had been ongoing since the beginning of the incident. Pirone's unlawful instruction to Grant, "You can't take pictures," was simply a continuation of his previous behavior. However, there was a marked change in the behavior of other police after the shot, when they attempted to confiscate cameras (GioSifaTaufa, 2009, 07:25).

The variable of **whether police made any attempt to prevent the recording** is positive. During the preliminary hearing held June 4, 2009, Judge Clay summarized:

> Mr. Grant had a cell phone camera in his hands which appeared from the video he was attempting to document the conduct of the BART officers. ...Officer Pirone who says, "You can't take pictures," and then subsequently the officer slammed Mr. Grant down on the ground... Within seconds, Oscar Grant is shot. (*California v. Mehserle*, 2009a, p. 1058).

The variable of **whether police made any attempt to acquire, confiscate, or destroy the video** is positive. Carazo testified that police tried to confiscate her camera, Cross testified that BART police did confiscate his camera's data card (Blanchard, 2009), and Vargas gave an interview in which she described how Domenici ordered Vargas to hand over her camera:

> I hopped back on the BART train. At this point, the female officer approaches me, the doors shut, and she's banging on the plastic, uh, of the door, telling me to give her my camera, and I tell her, No. (GioSifaTaufa, 2009, 07:25)

The variable of **whether the police gave any unlawful instruction to the videographer regarding the video** is positive. Vargas gave an interview in which she described how Domenici ordered Vargas to hand over her camera: "I hopped back on the BART train. At this point, the female officer approaches me, the doors shut, and she's banging on the plastic, uh, of the door, telling me to give her my camera, and I tell her, No" (GioSifaTaufa, 2009, 07:25). More importantly, in the preliminary hearing transcripts, Judge Clay summarized:

> Mr. Grant had a cell phone camera in his hands which appeared from the video he was attempting to document the conduct of the BART officers. ...Officer Pirone who says, "You can't take pictures," and then subsequently the officer slammed Mr. Grant down on the ground... Within seconds, Oscar Grant is shot. (*California v. Mehserle*, 2009a, p. 1058).

The variable of **whether police detained, cited, or arrested the videographer** is positive. As noted previously, Grant was detained, he attempted to record the police with his cell phone camera, and he was then shot by the police during his arrest (*California v. Mehserle*, 2009a, p. 1058). However, this variable is apparently negative for most of the videographers; although there was a report of other cameras being confiscated (Fox News, 2009), there was no report that the videographers were detained.

The variable of **when the video was available via the Internet** is no later than January 4, when YouTuber GioSifaTaufa uploaded the Vargas video and interview (GioSifaTaufa, 2009). An edited version of the interview was later broadcast on KTVU.

The variable of **when the video was available via broadcast news media** is no earlier than late Saturday, January 3, after Vargas had delivered a copy of her video to KPIX CBS-5 (GioSifaTaufa, 2009). It is not clear from available public documents whether CBS-5 broadcast the tape immediately; neither the station nor the network has maintained the video in any online archives, and some sources state that KTVU was first to broadcast a video of the incident (Stannard & Bulwa, 2009).

The variable of **when images from the video were available via print news media** is evidently January 7, in the *San Francisco*

Chronicle; prior to this, the newspaper included links to the television stations' online videos (Stannard & Bulwa, 2009). It is possible that the alternative press ran pictures prior to this date, but those publications are not readily accessible within the document search and acquisition procedures.

The variable of **whether more than one camera captured the event** is positive; beginning January 5, multiple civilian videos appeared in rapid succession on area television broadcasts and station websites, YouTube, and other media outlets (GioSifaTaufa, 2009; monkeyassj, 2009; ajajaj1, 2009; TheDirtyNews, 2009).

The variable of **whether official CCTV, dashboard, or other video captured the event** is positive. The BART platform CCTV cameras captured a partial view of the event, and the CD of that footage was introduced at the preliminary hearing as People's Exhibit 27 (*Los Angeles Times*, 2010; *California v. Mehserle*, 2009a, p. 1040). The variable of **whether police initially admitted to the existence or possession of video of the event** is negative; for two days, BART officials maintained that the platform surveillance cameras did not record (Tucker et al., 2009). The variable of **when did police admit to possession of video of the event** is that on January 3 BART revised its position, stating that the cameras did record but didn't show the incident (Tucker, 2009). The variable of **whether police released official video of the event** is positive; a CD of the CCTV recording was introduced at the preliminary hearing on June 4, 2009 as Exhibit 27 and made public during the trial (*California v. Mehserle*, 2009a, p. 1040; *Los Angeles Times*, 2010). The variable of **whether official video of the event was available via the Internet** is positive; the CCTV recording was introduced at the preliminary hearing on June 4, 2009 and made public during the trial, including being uploaded to multiple locations on the Internet (*Los Angeles Times*, 2010).

The variable of **who was credited as the source for each medium of release of the video** was rarely the videographer. KTVU superimposed its logo on Vargas' video, and many print and online media credited that station, even for stories interviewing Vargas about recording the video. Some outlets credited YouTube. However, the videos released through the courts were titled, subtitled, or captioned with the last name of the correct videographer (*Los Angeles Times*, 2010).

The variable of **whether there was any effort to restrict, remove or prosecute the release of the video** is negative; based on the document search and acquisition procedures, there is apparently no evidence of such efforts aside from the initial attempts to confiscate cameras by BART police. However, there was an official effort to prevent Sophina Mesa from testifying about her receipt of Grant's cell phone image of Mehserle shortly before the shooting, and to exclude the image (Bulwa, 2010 June 10).

The variable of the **initial official response to handling of the videographers** is that, as noted previously, police attempted to confiscate cameras and in some instances were successful (Blanchard, 2009). Vargas gave an interview in which she described how Domenici ordered Vargas to hand over her camera, and she refused (GioSifaTaufa, 2009, 07:25). Pirone also prevented Grant from recording police, shortly before Grant was shot (*California v. Mehserle*, 2009a, p. 1058).

The variable of the **final outcome regarding the videographers** is: Grant was shot and killed, and the videographers Karina Vargas, Tommy Cross, Daniel Liu (Liu Tong), Margarita Carazo, and Jamil Dewar were all called to testify on one or more occasions (Blanchard, 2009; *California v. Mehserle*, all dates).

The variable of the **sequence of official actions regarding the videographers** is that investigators interviewed the videographers in the course of the various investigations; the prosecution called them to testify during the preliminary hearings (Blanchard, 2009; *California v. Mehserle*, all dates); and Vargas, Cross, Liu, and Dewar were called again to testify during the second day of the trial in Los Angeles (Bulwa, 2010b).

The variable of the **compensation of the videographer** is unknown, based on the document search and acquisition procedures.

The variable of the **affiliation of the videographers** is that they are all civilians; none of them are journalists, activists, or affiliated with the police (Blanchard, 2009; Bulwa, 2010b).

The variable of **police misconduct** is positive; an unarmed civilian was shot to death by police, which a jury later found to be involuntary manslaughter (*California v. Mehserle,* 2009b, p. 1; Stannard & Bulwa, 2009; Winston, 2011), although Judge Clay concluded the preliminary hearings with the statement, "There's no doubt in my mind that Mr.

Mehserle intended to shoot Oscar Grant with a gun and not a Taser" (*California v. Mehserle*, 2009a, p. 1061).

The variable of the **initial response of the law enforcement agency regarding the officers involved** was that after the shooting, Mehserle was immediately taken from the incident site to BART headquarters at the Lake Merritt station (*California v. Mehserle*, 2010, February 19, p. 4). Mehserle told a friend, BART Officer Terry Foreman, that he thought Grant had a gun and was going for his pocket; Mehserle did not say to Foremen that he meant to use his TASER (*California v. Mehserle*, 2010, February 19, p. 13).

The variable of the **sequence of official actions regarding the officers involved** is complex. The department announced that the as-yet-unnamed officer had been suspended pending investigation (Tucker et al., 2009). DA representatives arrived at BART headquarters to interview Mehserle; through his attorneys, he invoked his right to remain silent. Mehserle also refused to provide a statement to the department. On January 7, Mehserle resigned rather than cooperate with the investigation (*California v. Mehserle*, 2010, February 19, p. 5).

The coroner performed an autopsy on Grant, and ruled the gunshot the cause of death. (*California v. Mehserle*, 2010, February 19, p. 5).

Protests occurred (*California v. Mehserle*, 2009b, pp. 14-17, 26), and elected officials issued statements about the incident. County Supervisor Keith Carson and Oakland City Councilwoman Desley Brooks both publicly used the term "execution" to describe what they saw on Vargas' video. Oakland Mayor Ron Dellums asked the court not to grant bail to prevent more riots, and issued a press release more than a week prematurely stating that Mehserle had been released. On January 8, BART's elected directors apologized to the victim's family. The elected BART Board of Directors issued public apologies for Mehserle's conduct, and called for the resignation of the BART police chief and BART's General Manager. In August, the BART police chief announced his retirement. State Assemblyman Sandre Swanson, chairman of the Black Legislative Caucus, and Attorney General Jerry Brown both publicly questioned why the DA was taking so long to file charges. Brown also announced that he was dispatching a high level aide to the DA's office to speed things along. Swanson and another legislator introduced a bill to create an oversight board for the BART police. Congresswoman Barbara Lee asked that local charges be filed, and also stated that if the DA did not prosecute, that she would ask for a Section 1983 prosecution (*California v. Mehserle*, 2009b, pp. 12-13).

On January 13, Mehserle was charged with one count of murder. Officials had to retrieve Mehserle from Nevada, where he had fled after resigning, reportedly because of death threats; he pled Not Guilty (Bulwa, et al., 2010).

BART hired the law firm Myers Nave to conduct an investigation of the incident. The investigation ran from February 11 through July 31, 2009, and resulted in a policy report critical of the department and of BART oversight (*California v. Mehserle*, 2010, February 19, p. 14).

At the preliminary hearings of May 27 and June 3 and 4, BART officers made statements that video evidence would contradict (Blanchard, 2009; *California v. Mehserle*, all dates). On June 16, the Alameda County district attorney filed a felony murder accusation against Mehserle, with a firearm specification, which under California law carries an additional three years' prison sentence. BART Police Chief Gee retired at the end of 2009 with full benefits.

By October, surveys showed that 97.7% of those polled in Alameda County were aware of the case, and 70% had already formed opinions (*California v. Mehserle*, 2009b, p. 6). On October 16, Mehserle's trial was ordered moved to Los Angeles (*California v. Mehserle*, 2009b).

BART fired Officer Marysol Domenici March 24th, 2010, and fired Officer Tony Pirone April 22nd, 2010 (Bulwa, 2010a; Bulwa, 2010d).

Mehserle's trial ran throughout June 2010, and included use-of-force expert analysis of the videos indicating that Mehserle had intended to use his TASER. On July 8th, Mehserle was convicted of involuntary manslaughter with a gun enhancement, which the judge set aside. He was sentenced to two years, with double credit for time already served.

The variable of the **final outcome regarding the officers involved** was that Mehserle was released on June 13th, 2011 after having served eleven months in Los Angeles County Jail. He was never placed in the state prison system, was in a private cell in protective custody, and was never in the general population (Winston, 2011). Domenici has since been rehired by BART, with back pay, after mediation required by the police union's contract with BART (Bulwa, 2010d).

For the variable of **final outcome of the law enforcement agency regarding policy changes, if any**: The department was subject to outside review by an independent law firm and by officials from the National Organization of Black Law Enforcement. Both organizations submitted public reports that criticized the department and its actions relating to the Grant shooting (*California v. Mehserle*, 2009b, pp. 25-26). BART thereafter hired an independent police auditor, Mark P. Smith; tripled the training hours for police; hired a new chief of police; began streaming board meetings live; and selected an 11-member Citizen Board (BART, 2011).

The variable of the **affiliation of the civilian subject** appears to be entirely civilian; he was not a journalist, nor was he affiliated with any media organization or with the police, or with any activist groups known for provoking the police (Blanchard, 2009; *California v. Mehserle*, 2009b, p. 12).

The variable of **the initial official response to handling of the civilian subjects** is complex, but is thoroughly described in Mehserle's trial brief. At 2 AM January 1 2009, BART police responded to a train operator's report of a fight on a train from San Francisco to the East Bay. The fight, between Grant and David Horowitch, had broken up by the time the train stopped at Platform 1 of the Fruitvale BART station, where Grant and about ten friends decided to get off. As they exited the train, Pirone walked toward them with his TASER in hand. Grant, Greer, and Dewar walked back onto the train. Pirone ordered Reyes, Jackie Bryson and Nigel Bryson to sit against a concrete wall; they complied. Pirone waited for his partner, Domenici, about two minutes. When she arrived, Pirone returned to the train, and pointed his TASER's laser sight in Grant's face to encourage him to comply with Pirone's order to get off the train. Pirone took Grant by the elbow and escorted him to the wall, where he had Grant sit next to his friends. Pirone then returned to the train, and began shouting profanely at Greer to get off the train. Pirone boarded the train, grabbed Greer, forced him onto the platform near the others, and used a leg sweep to drop Greer face-down on the platform. Pirone then handcuffed Greer (*California v. Mehserle*, 2010, February 19, p. 2).

At least two trains were at the platform, and were full of passengers due to the New Year's celebration. Pirone's profanity and excessive use of force drew a verbal response from the passengers, and more of them began recording video, including Cross, Liu, Carazo,

Dewar, and Vargas. Multiple civilian cameras captured the incident from a variety of angles and distances (*California v. Mehserle*, 2010, February 19, pp. 2-3, 6-10).

Grant, Reyes, and J. Bryson stood up and verbally protested Pirone's actions. Pirone grabbed Grant, shook him, delivered an elbow strike to Grant's head, then forced Grant to the ground. Pirone pointed his TASER at the others and ordered them to sit back down; they complied (*California v. Mehserle*, 2010, February 19, p. 3). Multiple witnesses testified that Pirone continued to use violent and profane language in dealing with the detainees throughout the incident, and that at no time during the incident did Grant appear to be resisting (*California v. Mehserle*, 2010, February 19, pp. 5-13).

Mehserle and Woffinden arrived on the platform, at which Pirone walked to the front of the train to speak with the operator (*California v. Mehserle*, 2010, February 19, pp. 3-4). At this time, Grant was evidently able to take a photograph of Mehserle with his cell phone camera, showing that Mehserle had his TASER out and pointed at Grant (Grant, 2009). Guerra arrived and stood with Mehserle and Woffinden in front of the five detainees. When Pirone returned, he told Mehserle and Guerra that Grant and J. Bryson were to be arrested. Grant rose to his feet, and Pirone told him to "Sit the fuck down" (*California v. Mehserle*, 2010, February 19, pp. 3) and forced Grant to a seated position, kneeing him in the face so his head struck the concrete wall. At this, J. Bryson stood up to protest; Mehserle forced him to a seated position. Per Pirone's orders, Mehserle and Guerra handcuffed J. Bryson, pulling him to his knees and directing him to place his hands behind his back; he complied (*California v. Mehserle*, 2010, February 19, pp. 3-4).

Mehserle then moved behind Grant, and brought him to a kneeling position; Grant complied. After Grant placed his hands behind his back, Mehserle forced him face down onto the platform, causing Grant to fall across Reyes' left leg. Pirone pushed down on Grant's head and shoulders while Mehserle pushed down on Grant's lower back (*California v. Mehserle*, 2010, February 19, p. 4).

Flores and Knudtson arrived on the platform, and Knudtson tackled Anicete, who had been verbally protesting the police actions. There were now seven police on the platform (*California v. Mehserle*, 2010, February 19, p. 4). Reyes attempted to tell police that Grant was on his leg. Mehserle drew his pistol. Pirone repositioned Grant, allowing Reyes to pull his left leg out from under Grant. Mehserle stood up and shot Grant once in the back (*California v. Mehserle*, 2010, February 19, p. 4).

Guerra retrieved a trauma kit from his patrol car and applied a bandage to the wound in Grant's back (*California v. Mehserle*, 2010, February 19, p. 4). Grant was transported to Highland hospital where he underwent surgery, and was pronounced dead later that morning (Tucker et al., 2009).

About two minutes after the shooting, Mehserle stated to Pirone, "Tony, I thought he was going for a gun" (*California v. Mehserle*, 2010, February 19, p. 4). Grant was unarmed during the incident (Stannard & Bulwa, 2009).

It is evident from the videos that police attempted to keep civilians, particularly those using cameras, at a distance from the five detainees (GioSifaTaufa, 2009; *Los Angeles Times*, 2010; monkeyassj, 2009; ajajaj1, 2009; TheDirtyNews, 2009). Immediately following the shot, police began confiscating cell phones and cameras from civilians; several escaped on departing trains (Blanchard, 2009; GioSifaTaufa, 2009, 07:25).

The variable of the **sequence of official actions regarding the civilian subjects** was complex. Initial official statements to the press were significantly at odds with the facts. BART spokesman Jim Allison stated that two groups of young men had been fighting on the train, and that police were trying to separate the groups as they continued to fight on the station platform (Tucker et al., 2009). Once the videos became available online, the official line shifted to assertions that a video can't show everything. However, six videos from widely divergent angles presented one consistent story (Stannard & Bulwa, 2009; GioSifaTaufa, 2009; *Los Angeles Times*, 2010; monkeyassj, 2009; ajajaj1, 2009; TheDirtyNews, 2009). Mehserle's defense attorneys consistently attempted to blame the civilians for the incident. However, the weight of testimony and evidence from the initial investigations through to the jury verdict made it clear that Mehserle and Pirone were at fault, that several police had perjured themselves repeatedly, and that police had exacerbated what should have been an exercise in keeping the peace with minimum use of force (*California v. Mehserle*, all dates).

The variable of the **compensation (if any) of the subject** is positive. The court approved a distribution of funds out of the $1.3 million settlement for Grant's estate of $685,375 to Grant's mother,

$457,067.72 to Burris for attorney's fees, $146,932.28 to Burris for costs, $5,625 to the California Victims' Compensation Board as repayment for assistance, and $5000 to Grant's estate, payable to Grant's daughter on her 18th birthday. The separate settlement for Grant's daughter was $1.5 million, structured so the total value of present and future payments was projected to exceed $5 million (*Johnson et al. v. Bay Area Rapid Transit et al.*, 2011, pp. 2, 8).

The variable of the **final outcome regarding the civilian subject** was that Grant's child and family were financially taken care of by the settlements (*Johnson et al. v. Bay Area Rapid Transit et al.*, 2011, pp. 2, 8).

The variable of **what third parties (if any) involved themselves in the legal case** is none. Officials of the local and state branches of the NAACP called for murder charges, and Amnesty International and the Meiklejohn Civil Liberties Institute pointed to Mehserle's actions as proof of racial profiling and police brutality (*California v. Mehserle*, 2009b, p. 13). However, Grant's relatives did not have the legal assistance of any other parties, and there were no *amicus* briefs filed in any of the criminal or civil cases.

The variable of ***cui bono?*, or who benefits?**, in this case was that Grant's child and family were financially taken care of by the settlements (*Johnson et al. v. Bay Area Rapid Transit et al.*, 2011, pp. 2, 8), and passengers and activists working for the reform of the BART police saw several changes (BART, 2011).

The variable of **how many times the video was viewed online** is that by January 7, 2009, the video had been downloaded from the website of KTVU Channel 2 more than 500,000 times (*California v. Mehserle*, 2009b, p. 5). YouTube has compiled 2,274 videos about the incident, with a total of nearly 12 million views (YouTube, 2011). The earliest three YouTube videos have accumulated 1 million, 590,000, and 1.2 million views, respectively (monkeyassj, 2009; ajajaj1, 2009; TheDirtyNews, 2009).

The variable of **how much press coverage the event received** was in the thousands of reports, the most of all the cases examined. A survey of potential jurors in Alameda County showed that an overwhelming majority, 97.7%, recognized the case (*California v. Mehserle*, 2009b, p. 6). A comprehensive assessment of early coverage appeared in an October 10 court order:

> Eighteen San Francisco Bay Area newspapers published 1,867 articles covering this case between January 1, and August 31, 2009. …the four largest newspapers in the Bay Area published an additional 70 articles between September 1, and October 7, 2009. Six local television stations, including the local ABC, CBS, Fox and NBC affiliates, broadcast 1,970 television news segments between January 1, and August 31, 2009. Three local stations aired 343 radio news stories between January 1, and May 18, 2009. (*California v. Mehserle*, 2009b, pp. 3-4)

The court agreed to change the venue of Mehserle's trial. However, the Bay Area press coverage continued at high levels throughout the trial. As of February 1 2012, there is ongoing coverage at a lower level, some in connection with the appeals to the civil suit rulings, but most of the coverage connects the Grant shooting to ongoing problems with BART and its police.

The variable of **how much Internet discussion was linked to this event** is that there were over 20,000 comments on the top three YouTube videos alone (monkeyassj, 2009; ajajaj1, 2009; TheDirtyNews, 2009). A Google search for +Mehserle +Grant retrieves over 52,000 blogs, nearly 18,000 forums, and over 4,500 videos. This case has generated far more Internet discussion than any of the other cases in this study.

CASE STUDY V: ANONYMOUS, MORALES, PIGOTT, 2008-09-24

On September 24, 2008, from a public sidewalk across from 489 Tompkins Avenue in the Bedford-Stuyvesant neighborhood of Brooklyn, New York, a civilian using a cell phone camera videorecorded NYPD Emergency Services Unit Lt. Michael Pigott ordering Police Officer Nicholas Marchesona to fire a TASER at psychiatric patient Iman Morales; Morales then fell to his death. This case is included for direct replication (similar results for expected reasons), per Yin (2003, pp. 5, 110). The phenomenon is a police-civilian interaction in American public space during which police misconduct may have occurred, which was videorecorded by a civilian, and the videorecording was distributed online.

Research Question and Hypotheses

For Research Question 1, what is the outcome of user-generated online video on police-civilian interactions in American public space?, the outcomes of this case include: retraining of all 440 NYPD ESU officers; assignment of a new ESU commanding officer; Pigott's suicide; Marchesona's clearing, return to duty, and promotion to detective; the termination of Morales' mother's lawsuit; and the dismissal of Pigott's widow's suit against the city (Gould & Gendar, 2008; Long, 2008, 2009; Pearson & Gendar, 2009 January 13; *Piggott v. City of New York et al.*, 2011; *Negron v. City of New York et al.*, 2011).

For this case, Hypothesis 1, that user-generated online video has the potential to improve accountability in police-civilian interactions in American public space, is evidently positive; the NYPD immediately began corrective action, and the senior officer apologized and committed suicide.

The value of Hypothesis 2, that user-generated online video is significantly different from professional video journalism in its effects on accountability in police-civilian interactions in American public space, in this case is evidently positive. Even though the incident had gone on for some time, none of the New York television stations had judged the story important enough to send a camera crew until after Morales had died (NY1, 2008). It was a civilian cell phone camera that recorded the incident, and the video's first availability was through the website of a newspaper rather than a television station (Doyle, et al., 2008).

Hypothesis 3, that there are strong motivations for police to continue to attempt to restrict civilians' First Amendment rights to photograph police in public spaces, is evidently positive; in this case, a user-generated online video not only ended a police lieutenant's career, but also led to a civil suit against his estate (*Negron v. City of New York et al.*, all dates).

Documents

Documents for this case include over 90 wire service news stories from LexisNexis Academic News alone, plus mentions within stories on TASER use and police misconduct. There are over 60 court documents, but relatively few are crucial; in particular, the official documents appended to court filings provide the most useful details (*Piggott v. City of New York et al.*, all dates; *Negron v. City of New York et al.*, all dates). Pigott's suicide note was made public by his widow. There has been one law review article that cites the case. There are over 100 data files, including 16 media clips from YouTube, broadcast television, and other websites. There are hundreds of audience comments on YouTube and media websites; a very large number of posts on a wide range of blogs; a Wikipedia page; a Facebook page; and numerous letters to the editor and op-eds in the online editions of newspapers. A Google search for +Morales +Pigott +Taser yields 681 unique results, and estimates 22,000 without removing duplicates.

Relevant Variables of Interest

The variable of **whether the police were aware of the camera** is positive, based on Marchesona's statement that "the crowd in turn was telling Mr. Morales to jump as they took photos with their camera phones" (*Negron v. City of New York et al.*, 2010 December 20, 54, p. 37). Therefore, the variable of **when the police became aware of the camera** is before the beginning of the video. The variable of **whether there was any marked change in police behavior once they were aware of the camera** is evidently negative, based on the video (ChooseRonPaul, 2008). The variables of **whether police made any attempt to prevent the recording, whether police made any attempt to acquire, confiscate, or destroy the video, whether police detained, cited, or arrested the videographer**, and **whether the police gave any unlawful instruction to the videographer regarding the video** are unknown, based on the document search and acquisition procedures.

The variable of **when the video was available via the Internet** is September 24, 2008; the website of the *New York Post* was the origin, followed within 24 hours by multiple duplicates taken from the *Post* and uploaded to YouTube (Doyle, et al., 2008; ChooseRonPaul, 2008). The variable of **when the video was available via broadcast news media** is approximately five hours after the incident, when WNBC broadcast the *Post* video at 7 p.m. September 24, 2008; NY1 and other stations followed quickly (WNBC, 2008; NY1, 2008; CW11, 2008). The variable of **when images from the video were available via print news media** is September 25, 2008, in the *New York Post* print edition

(Doyle, et al., 2008). The variable of **whether more than one camera captured the event** is positive; several amateur and professional photographers captured still images of the event, some of which were edited into the *Post* video (Doyle, et al., 2008; WNBC, 2008; Shapiro, 2008; Hutton, 2008), and CW11's evening broadcast included some of Racquel McDonald's cell phone video of an earlier part of the incident, but not the TASER firing or Morales' fall (CW11, 2008).

The variable of **whether official CCTV, dashboard, or other video captured the event** is apparently negative, based on the document search and acquisition procedures, as no such videos have appeared as public documents, nor were any mentioned in discovery proceedings in either of the two civil lawsuits resulting from the incident (*Negron v. City of New York et al.*; *Piggott v. City of New York et al.*). For the same reasons, the variables of **whether police initially admitted to the existence or possession of video of the event**, **when did police admit to possession of video of the event**, **whether police released official video of the event**, and **whether official video of the event was available via the Internet** are not applicable.

The variable of **who was credited as the source for each medium of release of the video** is "amateur video" for the WNBC and NY1 broadcasts (WNBC, 2008; NY1, 2008); the *Post* did not initially credit any source for the video. CW11's evening broadcast credited Racquel McDonald for her cell phone video of an earlier part of the incident (CW11, 2008).

The variable of **whether there was any effort to restrict, remove or prosecute the release of the video** is negative; based on the document search and acquisition procedures, there is no evidence of any such effort. The variables of the **initial official response to handling of the videographer** and the **sequence of official actions regarding the videographer** are unknown, because the videographer has not been identified.

The variable of the **final outcome regarding the videographer** is that the *New York Post* has not disclosed any amount paid for the video, nor the videographer's name; because both civil suits ended before trial, the videographer was not called to testify, and so remains anonymous (*Negron v. City of New York et al.*; *Piggott v. City of New York et al.*). The variable of the **compensation of the videographer** remains undisclosed. The variable of the **affiliation of the videographer** has been reported as amateur civilian, but that cannot be confirmed within the document search and acquisition procedures. Because the *New York Post* edited the original video before uploading it, metadata that might identify the videographer are not present in any of the online video files.

The variable of **police misconduct** is positive; Pigott's order to Marchesona was not in line with a 10-page interim order, issued three months previously, for the use of a "Conducted Energy Device (CED), commonly known as a Taser…when possible, the CED should not be used…in situations where the subject may fall from an elevated surface" (Browne, 2008 September 25).

The variable of the **initial response of the law enforcement agency regarding the officers involved** was, to question all officers on the scene other than Pigott and Marchesona, then to place Marchesona on desk duty, and to strip Pigott of his gun and badge and assign him to answer telephones at the motor pool pending outcome of the investigation (Barish, 2008; Long, 2008; Leavitt, 2009). The Brooklyn District Attorney asked that neither Pigott nor Marchesona be interviewed by the NYPD as the investigation continued (Browne, 2008 September 25; Faheem & Hauser, 2008).

Following this, the variable of the **sequence of official actions regarding the officers involved** was, the police commissioner and the police spokesman made statements to the public and the press that Pigott made a "mistake," and that "The order to employ the Taser … appears to have violated guidelines" (Browne, 2008). Department officials reportedly told the media and Pigott that his actions were "improper," that his job was in question, and that he might be facing criminal charges (Leavitt, 2009). Officials also reportedly told Pigott he would never work in the ESU again (Ginsberg, 2009). City attorneys reportedly informed Pigott that "they might not defend or indemnify him in a civil suit," contrary to standard police practices (Leavitt, 2009).

Pigott returned to the ESU facility, cut the lock off another officer's locker, and used the firearm he found there to shoot himself in the head. Other ESU officers discovered his body, including his suicide note, which read:

> Dear Sue, Rob, Mikey and Liz, I love you all. I am sorry for the Mess! I was trying to protect my guys that day! I ordered

> Nick Marchesona to fire the Taser! I can't bear to lose my family and go to jail. (Leavitt, 2009)

The police initially suppressed the note; it was made public by Pigott's widow as part of her lawsuit against the city (*Piggott v. City of New York et al.*).

Following Pigott's suicide, the department and the city retreated from their earlier positions. Commissioner Kelly persuaded the city's attorneys to represent Pigott's estate in any civil action. However, no ranking police official attended Pigott's funeral (Leavitt, 2009; *Piggott v. City of New York et al.*). Officials also made apparently contradictory statements to the press, including that Pigott had been assured that he would not face criminal charges (Parascandola, 2008).

NYPD Internal Affairs Bureau interviewed Marchesona and obtained a statement October 30, 2008, which was later introduced in court filings and became public (*Negron v. City of New York et al.*, 2010 December 20, 54, pp. 35-38).

Prior to the incident, Marchesona had been scheduled for promotion to Detective. That promotion was made official five weeks after the incident (Pearson & Gendar, 2008). Following the internal investigation, Marchesona was cleared and returned to duty (Long, 2009). Marchesona was also defended by city attorneys against Negron's civil suit (*Negron v. City of New York et al.*, all dates).

The variable of the **final outcome regarding the officers involved** is that Pigott committed suicide October 2, 2008; that Marchesona was cleared and restored to duty, with no delay of his scheduled promotion to detective; and that the lawsuit against Pigott's estate was terminated in September, 2011 (Long, 2008, 2009; *Negron v. City of New York et al.*, 2011).

For the variable of **final outcome of the law enforcement agency regarding policy changes, if any**: Commissioner Kelly assigned the ESU a new commanding officer, Deputy Chief James Molloy, and the Monday following the incident, all 440 officers of the ESU attended eight hours of mandatory retraining in dealing with emotionally disturbed persons (Gould & Gendar, 2008). Kelly also stated that, rather than change NYPD training, officers just needed to be reminded of the approved tactics and procedures (Gendar, 2008 September 30).

The variable of the **affiliation of the civilian subject** appears to be entirely civilian; he was not a journalist, nor affiliated with any media organization or with the police, or with any activist groups known for provoking the police. However, Morales was also a special case: a psychiatric patient on medication to control psychotic symptoms, paranoid delusions, and depression, who had hepatitis C, and who had, one week before the incident, been informed that he was also positive for HIV (*Negron v. City of New York et al.*, 2010 November 20, p. 20; 2010 December 20, 54, p. 40).

The variable of **the initial official response to handling of the civilian subject** was to initiate an internal investigation, to place Marchesona on modified duty, and to strip Pigott of his gun and badge and assign him to answering phones in the motor pool (Chung, 2008 October 3; Gendar, 2008; Gould & Gendar, 2008; Long, 2008 October 3; Leavitt, 2009).

The variable of the **sequence of official actions regarding the civilian subject** was: On September 24, 2008, NYPD 911 services received a call from Negron, Morales' mother, requesting assistance (Chung, 2008; *Negron v. City of New York et al.*, 2010 December 20, 55, p. 9). The local precinct requested ESU assistance with an EDP (emotionally disturbed person). Pigott and Marchesona were part of the ESU response, and arrived at 1:52 p.m. ESU received information from Negron that Morales had been diagnosed with HIV, had been off his psychiatric medications for a week, and had made suicidal statements earlier that day. Marchesona responded to Morales' leaning out a third-floor window and yelling at the crowd, "You're going to kill me…I'm going to take everyone with me…I'm going to die...you're all going to die with me" (Doyle et al., 2008) by going to Morales' third-floor apartment with Detective McLaughlin. Marchesona then responded to a fourth-floor tenant's call for help, that Morales was trying to break into her apartment from the fire escape. Marchesona saw Morales trying to pull the apartment's air conditioner loose to gain entrance to the tenant's apartment. Police in Morales' apartment requested that Morales come back inside the building. More ESU personnel and equipment arrived, including a safety harness used to secure police to the structure while dealing with an EDP. With Morales on the fire escape, and the crowd, estimated at 200 persons, encouraging him to jump, the ESU reclassified the incident from a "barricaded EDP job" to a "jumper job". An ESU truck with an airbag had also been summoned, but was not yet on the scene. ESU officers continued to talk to Morales from his third-floor apartment window. Two ESU officers moved down the fire escape toward Morales, who climbed onto the adjacent fire ladder, saying, "Oh, I'm so going to die today" (*Negron v. City of New York et al.*, 2010 December 20, 54, p. 37). Marchesona left the building, at which point Pigott confirmed that Marchesona had a TASER. Marchesona saw two ESU officers in the process of harnessing themselves to the fire escape so as to pull in Morales. However, the officers were not yet secure, were in danger of falling (Del Signore, 2008), and Morales was hitting them with a fluorescent light tube from where he stood on top of a roll-down security gate housing. Pigott ordered Marchesona to "Get over here and Taser him" (*Negron v. City of New York et al.*, 2010 December 20, 54, p. 37), which Marchesona did. Marchesona reported that he called to the harnessed officers to grab Morales. They reported that they did not hear him; no other police intervened as Morales fell from the security gate and landed on his head on the sidewalk ten feet below, at approximately 2:27 p.m. (Faheem & Hauser, 2008). Morales was transported to Kings County hospital and pronounced dead at 2:34 p.m. (*Negron v. City of New York et al.*, 2010 November 10, p. 6).

The variable of the **compensation (if any) of the subject** is none; the suit brought for $10 million by Negron was terminated in September, 2011 (*Negron v. City of New York et al.*). The variable of the **final outcome regarding the civilian subject** is that his funeral was October 2, 2008 (Del Signore, 2008).

The variable of **what third parties (if any) involved themselves in the legal case** is none; although community activists and organizations including the New York Civil Liberties Union, the Justice Committee, and the National Latino Officers Association spoke to the media about this case, none of them filed briefs or otherwise participated in *Negron v. City of New York et al.* Similarly, although the Lieutenants Benevolent Association and Police Organization Providing Peer Assistance (POPPA) spoke to the media about this case, none of them filed briefs or otherwise participated in *Piggott v. City of New York et al.* (Leavitt, 2009; Kolodner, 2008; Long, 2008; Colangelo, 2008; *Negron v. City of New York et al.*; *Piggott v. City of New York et al.*).

The variable of ***cui bono?*, or who benefits?**, for this case is any civilian, and particularly any emotionally distressed person, who might encounter NYPD ESU police who are equipped with TASERs.

The variable of **how many times the video was viewed online** is over 117,000 for the most popular copy posted to YouTube; however, the original *New York Post* site does not publicly report viewer statistics, and most of the other media reports published links to the *Post* site (ChooseRonPaul, 2008; Doyle, et al., 2008).

The variable of **how much press coverage the event received** includes over 90 wire service news articles from LexisNexis Academic News alone, and 16 media clips from YouTube, broadcast television, and other websites. Although the NYC area coverage has been heaviest, the national networks and wire services also reported the story. The tabloid media in particular have covered this story extensively.

The variable of **how much Internet discussion was linked to this event** includes over 450 viewer comments on the YouTube video; a very large number of posts on a wide range of blogs; a Wikipedia page, "Iman Morales Taser incident"; a Facebook page; multiple reader comments on news websites; and numerous letters to the editor and op-eds in the online editions of newspapers. A Google search for +Morales +Pigott +Taser yields 681 unique results, and estimates another 22,000 without removing duplicates.

CASE STUDY VI: MORRIS, MONETTI, COBANE, 2010-04-17

At 1:30 a.m. on April 17, 2010, on Westlake Avenue in Seattle, Judson Morris videorecorded Seattle PD Detective Shandy Cobane kicking Martin Monetti in the head, and Officer Mary Woollum stomping on Monetti's leg. This case is included for direct replication (similar results for expected reasons), per Yin (2003, pp. 5, 110); the phenomenon is a police-civilian interaction in American public space during which police misconduct may have occurred, and of which user-generated video was posted online.

Research Question and Hypotheses

For Research Question 1, what is the outcome of user-generated online video on police-civilian interactions in American public space?, the outcomes of this case include, for the police: Cobane apologized (MoxNewsDotCom, 2010; KIRO-TV, 2010c), was suspended for 30 days without pay, and was demoted to patrol officer; Woollum was given a written reprimand; both Cobane and Woollum were reassigned;

and the incident's supervising officer, Keith Swank, was suspended for 10 days without pay and ordered to undergo retraining (Miletich, 2011). For the journalists: a news director resigned and a senior assignment editor was fired (O'Hagan, 2010b); the videographer was fired (KIRO-TV, 2010b; O'Hagan, 2010a); the videographer was charged with possession of stolen camera gear (McNerthney, 2010b; KIRO-TV, 2010v; Q13 Fox News, 2011); Monetti filed a civil rights suit (*Monetti v. City of Seattle, et al.*, 2011) which was settled for $150,000 (Green, 2012, June 27); and the Justice Department carried out a civil-rights probe of the Seattle PD, and mandated policy changes (KIRO-TV, 2010j; McNerthney, 2011; U.S. Department of Justice, 2011c). As of September 2012, the Seattle PD was operating under a court-appointed monitor (*United States v. City of Seattle,* 2012, September 21).

For this case, Hypothesis 1, that user-generated online video has the potential to improve accountability in police-civilian interactions in American public space, in this case is evidently positive. The publication of the video prompted public statements by police and personnel actions by the management of a television station, and contributed to a federal investigation.

The value of Hypothesis 2, that user-generated online video is significantly different from professional video journalism in its effects on accountability in police-civilian interactions in American public space, in this case is also positive. Professional television journalists refused to distribute the video (KIRO-TV, 2010b), and its YouTube distribution bypassed that mediation (Morris, 2010).

Hypothesis 3, that there are strong motivations for police to continue to attempt to restrict civilians' First Amendment rights to photograph police in public spaces, is evidently positive in this case. User-generated online video was instrumental in damaging several police careers (Miletich, 2011) and in forcing reforms of the department (US DOJ, 2011c; *United States v. City of Seattle,* 2012, September 21).

Documents

The case data exceed 160 files, including 17 media clips from YouTube, broadcast TV, radio, and Seattle PD and city official releases. There were over 62 court documents in *Monetti v. City of*

Seattle, et al., and another 13 for *United States v. City of Seattle*. There were over 70 news wire articles from LexisNexis Academic News alone. Seattle area media have reported the case extensively, connecting it with similar incidents; significant developments also made the national news. The incident video (Morris, 2010) was shot by a freelance videographer recently employed by Seattle broadcast station KCPQ, also known as Q13. The exact timing of distribution is a matter of some dispute, but the video eventually ended up on YouTube, then broadcast on KIRO-TV (Halsne, 2010b), then was widely reported. The videographer was interviewed at length by KIRO-TV, which hosts the uncut videos on their website. The apparent withholding of the video was reported internationally. KCPQ, the station that originally refused to air the video, announced that those responsible had resigned or been fired (O'Hagan, 2010b). The Seattle PD has made several public statements and press releases, which are available in full text through the department website. The U.S. Department of Justice has released a report and supplementary documents from its civil rights investigation. Cobane and Woollum neglected to report the use of force in the incident, which was the fact that led to disciplinary actions (Miletich, 2011); their original report has been made public in redacted form. Four other officers were cleared of misconduct by the SPD's Office of Professional Accountability (OPA). The Seattle City Attorney announced that Cobane and Woollum were not going to be charged with a hate crime, or with misdemeanor assault in the fourth degree. The civil rights suit, *Monetti v. City of Seattle, et al*, was settled in June, 2012; there were 62 documents filed prior to settlement. Monetti and his family held a press conference when they filed the suit; the video is available online. A subpoena (later made public) was served on the videographer for the original media card holding the video recording, and he attended a closed evidence hearing with his attorney, Seattle/King County NAACP president James Bible.

Relevant Variables of Interest

The variable of **whether the police were aware of the camera** is positive. It is evident from the video that police were not aware of the camera for the first two minutes of the video. Police behavior changed markedly when they realized they were being recorded, particularly in how much more politely they treated Monetti (Morris, 2010, 02:00+).

The variable of **when the police became aware of the camera** is approximately two minutes into the video (Morris, 2010, 02:00). The variable of **whether there was any marked change in police behavior once they were aware of the camera** is positive. Police behavior changed markedly when they realized they were being recorded, particularly in how much more politely they treated Monetti (Morris, 2010).

The variable of **whether police made any attempt to prevent the recording** is negative. Police evidently did not make any attempt to prevent the recording, and did not approach or interact with the videographer (Morris, 2010). For this reason, the variables of **whether police made any attempt to acquire, confiscate, or destroy the video, whether the police gave any unlawful instruction to the videographer regarding the video**, and **whether police detained, cited, or arrested the videographer** are also negative.

The variable of **when the video was available via the Internet** is that the videographer originally uploaded the video to YouTube on April 20 or 21 (McNerthney, 2010a). The videographer then sold exclusive broadcast rights to the video to KIRO-TV (KIRO-TV, 2010b, 05:00), and pulled the YouTube video (McNerthney, 2010a). An edited version of the video was available on May 6, 2010 via the KIRO-TV website. The videographer uploaded the unedited 7:33 length video to his YouTube channel, stringer253, on May 29, 2010 (Morris, 2010). The variable of **when the video was available via broadcast news media** is that the video was broadcast on May 6, 2010 during the 11 p.m. news on KIRO-TV (Halsne, 2010b). The variable of **when images from the video were available via print news media** is that images from the video were published May 6 in the *Seattle Post-Intelligencer*, an online-only Hearst newspaper (McNerthney, 2010a), but that the document search and acquisition procedures did not result in any evidence of the images appearing in print before that date.

The variable of **whether more than one camera captured the event** is apparently negative, based on the document search and acquisition procedures, as no other civilian videos have been made public at this time.

The variable of **whether official CCTV, dashboard, or other video captured the event** is positive, despite statements to the contrary by police. Eric Rachner successfully sued Seattle police over dashcam video they refuse to admit exists; he has since created the Seattle Police

Video Project website to access the dashcam database (Carter, 2011). A search of this database shows that Woollum, badge #6269, logged a dashcam video at 1:20 AM and another at 1:41 AM on April 17, bracketing the time of the incident (http://seattlepolicevideo.com/). Rachner has paid particular attention to these videos:

> Rachner says he can show that the department did not turn over every video from incidents apparently being reviewed by the DOJ. An example, he said, is the incident in which an officer threatened to beat the "Mexican piss" out of a suspect. Rachner said the DOJ got one department video of that incident, but the logs show that there are six other videos from that incident that were not turned over. (Carter, 2011)

It is worth noting that while all patrol vehicles were equipped with dashcams, the gang unit vehicles were not; Cobane's demotion to patrol officer will put him back in the view of department dashcams (KIRO-TV, 2011d).

For the previously stated reasons, the variables of whether police initially admitted to the existence or possession of video of the event is negative, the variable of when did police admit to possession of video of the event is not as of February 2, 2012, the variable of whether police released official video of the event is negative, and the variable of whether official video of the event was available via the Internet is negative (Carter, 2011).

The variable of **who was credited as the source for each medium of release of the video** is that Jud Morris was rarely credited for any version of the video. KIRO-TV, KCPQ/Q13, YouTube, no credit at all, or 'freelance videographer' were the most common credits.

The variable of **whether there was any effort to restrict, remove or prosecute the release of the video** is positive. KCPQ threatened legal action against KIRO-TV and the videographer, but no civil actions were filed, and the videos have remained available on the KIRO-TV website and on YouTube (O'Hagan, 2010a; Morris, 2010; Halsne, 2010b).

The variable of the **initial official response to handling of the videographer** is that around 1:30 a.m. on April 17, 2010, Judson Morris videorecorded the incident near 1264 Westlake Avenue in Seattle. The videographer talked with Monetti, recording a close-up of the bloody marks on Monetti's forehead. The videographer took the video to KCPQ later that day, Saturday morning, and showed it to senior assignment editor Cheri Mossburg and news director Steve Kraycik (KIRO-TV, 2010b, 01:20). According to the videographer, Mossburg called the police to discuss the video while they were watching it, which the videographer found odd: "It's not something you want to leak to the police before you have even finished watching the video" (O'Hagan, 2010a). The following Monday, the videographer was informed that the station was definitely not going to air the video (KIRO-TV, 2010b, 01:30).

The variable of the **final outcome regarding the videographer** is ambiguous. He was still working as a freelance videographer in the Seattle area as of February 2012, but since then his website domain registration has expired. In addition, the prosecution of the videographer for possession of stolen KCPQ cameras evidently concluded with his pleading guilty on September 9, 2011 to first degree possession of stolen property, and being placed in the Veteran's Drug Court program. The charges were to be dropped if he completed the one-year program (Q13 Fox News, 2011).

The variable of the **sequence of official actions regarding the videographer** is complex. The morning of Monday, April 19, the videographer was informed that the station was definitely not going to air the video (KIRO-TV, 2010b, 01:30). That evening, the videographer discussed the video with a producer and two writers in the KCPQ newsroom, who the videographer says encouraged him to put it on YouTube (KIRO-TV, 2010b, 02:25). The videographer uploaded the video to YouTube April 20 or 21; police became aware of the video, and informed KCPQ staff (O'Hagan, 2010a; McNerthney, 2010a). On Monday, April 26, the videographer received a phone call from KCPQ saying, "We heard you put this up on YouTube. Don't bother coming in tonight" (KIRO-TV, 2010b, 02:40; McNerthney, 2010a). The videographer then sold exclusive broadcast rights to the video to KIRO-TV (KIRO-TV, 2010b, 05:00), and pulled the YouTube video (McNerthney, 2010a).

After KIRO-TV broadcast the edited video on May 6 (Halsne, 2010b), police internal affairs attempted to contact the videographer to get the original recording. When there was a delay due to the videographer's nighttime work habits, internal affairs reportedly copied the YouTube video to forward to another investigator (KIRO-TV, 2010b, 07:00).

The videographer was subpoenaed on July 1, 2010, to give evidence and to hand over the original media card to the court on July 6. Although reporters were at the courthouse, the hearing was closed and there was a gag order on all participants. Reporters later concluded that since the videographer's name did not appear on jail rosters he must have complied with the court's orders (KIRO-TV, 2010u).

The videographer was later charged with possession of stolen KCPQ cameras; former news director Kraycik denied that he had agreed to sell the videographer the cameras (McNerthney, 2010b; KIRO-TV, 2010v).

The variable of the **compensation of the videographer** is that the videographer sold the video to KIRO-TV for the standard fee of $100 (KIRO-TV, 2010b, 05:00).

The variable of the **affiliation of the videographer** is that he was a freelance videographer or 'stringer' who had "a temporary full-time freelance gig with" (O'Hagan, 2010a) Seattle broadcast station KCPQ at the time of the incident. According to Morris, he is a self-taught videographer who had not used a video camera before buying one at a pawnshop about a year before the incident (O'Hagan, 2010a). At the time of the incident, the videographer claims he was off the clock, using his own equipment and his own vehicle (KIRO-TV, 2010b; O'Hagan, 2010a).

The variable of **police misconduct** is positive. Police misconduct was recorded, as evidenced by the ensuing investigations (both internal and external) and administrative punishments (KIRO-TV, 2010j; McNerthney, 2011; Miletich, 2011).

The variable of the **initial response of the law enforcement agency regarding the officers involved** was that an internal affairs investigation had begun the day of the incident, but the chief of police did not inform the mayor or the public until the video was broadcast (KIRO-TV, 2010e). It is evident that the investigation had begun before the second incident involving Cobane had occurred. The chief also confirmed that he was personally acquainted with Cobane, who had been with the department 16 years (KIRO-TV, 2010e). The police were evidently aware of the YouTube video shortly after it was uploaded on April 20 or 21, because they informed KCPQ about it sometime within the next five days (McNerthney, 2010a).

The variable of the **sequence of official actions regarding the officers involved** is complex. The department administratively reassigned Cobane and Woollum after KIRO-TV broadcast the video (KIRO-TV, 2010s). The administrative process was placed on hold, according to the acting chief of police, when the criminal case file was turned over to the county prosecutor in mid-May. At the same time, it became public that internal affairs had collected audio and video recordings of the incident from a number of patrol cars. A separate investigation was also in process to determine if any police had exerted pressure on KCPQ to prevent the video from being broadcast (KIRO-TV, 2010s). Later that month, the department suspended Cobane for 30 days without pay, citing the racial slur and other misconduct, which apparently included an April 24 incident that was also captured on video (Miletich, 2011; KIRO-TV, 2011a).

On September 1, the county prosecutor released a statement that Cobane would not be charged with a hate crime. Cobane did use "patently offensive language referencing the suspect's ethnicity" but did not direct a threat or assault toward Monetti specifically because of the detainee's race, according to the statement (KIRO-TV, 2010w).

The city attorney's office also reviewed the incident to determine whether either Cobane or Woollum could be charged with misdemeanor fourth-degree assault. Criminal Division Chief Craig Sims reviewed the case personally, and concluded that the use of force was lawful, but that "Cobane's use of a racial slur was not a necessary verbal tactic" (Holmes, 2010).

An external reviewer, Detective Gregory McKnight of the LAPD, submitted a letter on December 14, 2010 in reference to the Use of Force incident. Although his conclusion was that there was insufficient evidence for a criminal prosecution, McKnight pointed out significant gaps in the case file, particularly that statements from the officers involved were missing, that the source of Monetti's injuries and his allegations of being previously struck by Cobane were never addressed, and that the police did not follow best practices in their handling of the detainees. It is notable that McKnight was not provided with all the available recordings of the incident; the only video he reviewed was that recorded by the civilian videographer (McKnight, 2010).

The variable of the **final outcome regarding the officers involved** was administrative penalties including demotion, retraining, and suspension without pay, followed by return to duty; no police were fired. Cobane was suspended for 30 days without pay, taken off the gang unit, required to take part in community outreach training, and was demoted to patrol officer. Woollum was given a written reprimand, the officers were reassigned, and the incident's supervising officer, Keith Swank, was suspended for 10 days without pay and ordered to undergo retraining. The other four officers were cleared (Miletich, 2011; KIRO-TV, 2011k). KIRO-TV later reported that Cobane had been permitted to earn 167 hours of overtime during 2011, which reportedly "allowed him to evade a significant loss in overall pay" (Halsne, 2012).

For the variable of **final outcome of the law enforcement agency regarding policy changes, if any**: The chief of police announced that, "There'll be a presumption of termination in the future for anybody who uses racial or ethnic slurs" (KIRO-TV, 2011k).

In response to Rachner's setting up of seattlepolicevideo.com and his ongoing lawsuit against the department for its video policies and practices, a police spokesman stated that the department "applauds" Rachner for his work, and that "We believe that all information, once it has served its law-enforcement function, should be public. I think we've already seen that with this lawsuit and others that there is going to be some growth in the department over this issue" (Carter, 2011). In late December 2011, the OPA released an audit report that concluded that police were not using their dashcams according to guidelines, including not recording 23 percent of traffic stops. The report called for reminders to police and supervisors about recording policies, and recommended that a working group be established; the chief was reportedly already implementing those recommendations (Rosenthal, 2011).

The U.S. Department of Justice carried out a civil rights probe of the Seattle PD (McNerthney, 2011; KIRO-TV, 2011k; US DOJ, 2011a). In its findings letter, the DOJ stated:

> We also reviewed the video of the notorious incidents involving an officer's threat to "beat the f'ing Mexican piss" out of a suspect. It is troubling that the use of this racial epithet failed to provoke any of the surrounding officers to react, suggesting a department culture that tolerates this kind of abuse. Of greatest concern, neither of the two supervisors present admonished the officer at the scene. Nor did anyone

> report the incident to OPA until a third-party video of the incident was posted publicly. The number of people present, the failure to correct the officer, and the failure to immediately report the conduct all could be seen as a reflection of a hardened culture of accepting racially charged language. (2011b, p. 27)

The letter also includes the specific recommendation, "SPD and OPA should ensure that all in car video recordings are made available to supervisors for review" (p. 39). In the press release accompanying the report, the DOJ also stated, "Resolution of these findings will require a written, court-enforceable agreement that sets forth remedial measures to be taken within a fixed period of time" (US DOJ, 2011c). That agreement took the form of SPD operating under a court-appointed monitor as of September 2012 (*United States v. City of Seattle,* 2012, September 21).

The variable of the **affiliation of the civilian subject** appears to be entirely civilian; he is not a journalist, nor was he affiliated with any media organization or with the police, or with any activist groups known for provoking the police; his and his family's response to the incident was to ask the media to respect their privacy (KIRO-TV, 2010p).

The variable of **the initial official response to handling of the civilian subjects** is that at 1:30 a.m. on April 17, 2010, near 1264 Westlake Avenue in Seattle, Detective Shandy Cobane kicked prone detainee Martin Monetti in the head and threatened to "beat the fucking Mexican piss out of you" (Morris, 2010, 00:30; Halsne, 2010b). Officer Mary Woollum stomped on Monetti's leg. When Monetti was identified as an innocent bystander, police let him go. Morris talked with Monetti, recording the bloody marks on Monetti's forehead (Morris, 2010, 7:20).

The variable of the **sequence of official actions regarding the civilian subjects** was that Monetti received a public apology from the police at a press conference on May 7, the day after the video was first broadcast (KIRO-TV, 2010c). Videos of Cobane's apology became popular on YouTube (MoxNewsDotCom, 2010). The chief, who was out of the city for that week, also made a separate apology to Monetti at a news conference on his return (KIRO-TV, 2010o). Nearly a year later, Monetti's attorney filed a tort claim for $750,000 (KIRO-TV,

2011h). In response, the department publicly stated that Cobane had already received "the most severe discipline available short of termination" (KIRO-TV, 2011l). Monetti filed suit June 22, 2011 in U.S. District Court (*Monetti v City of Seattle et al*, 2011a), and the case was in the discovery phase as of February 2, 2012 (*Monetti v City of Seattle et al*, 2011d).

The variable of the **compensation (if any) of the subject** is that Monetti filed a civil rights suit (*Monetti v City of Seattle et al*, all dates), which was settled for $150,000 shortly after the court ordered the case to trial (Green, 2012 June 27).

The variable of the **final outcome regarding the civilian subject** was that Monetti received an apology from the department and from Cobane (Q13 Fox News, 2010 May 7), and the city settled Monetti's civil rights suit (Green, 2012 June 27).

The variable of **what third parties (if any) involved themselves in the legal case** is that the ACLU, NAACP, and El Centro de la Raza all issued public statements in support of Monetti, but none filed an *amicus* brief or otherwise overtly contributed to his civil rights suit (KIRO-TV, 2010f, 2010k; Narayan, 2010).

The variable of ***cui bono?*, or who benefits?**, is in this case, the civilian subject in particular, and civilians in the Seattle area generally, as their police are going through a major reform process, especially in transparency and in the use of video to document police-civilian interactions (Carter, 2011; US DOJ, 2011c, *United States v. City of Seattle,* 2012, September 21).

The variable of **how many times the video was viewed online** is only 2,500 views for the 7:33 unedited version, which was uploaded to YouTube May 29, 2010, long after the peak of the video's viral activity. However, there have been multiple edited versions of the video available since May 6, 2010, on television station websites and on YouTube. Several of the many YouTube copies have in excess of 100,000 views each (Morris, 2010; Halsne, 2010b), so the aggregate count is no less than 300,000 views.

The variable of **how much press coverage the event received** included over 70 news wire articles from LexisNexis Academic News. The Google News search +Cobane +police +video retrieves 61 articles, including archives. Local media covered the story intensively and in combination with ongoing issues of police accountability; Spanish-language radio was very active (KIRO-TV, 2010g). The *Seattle Times*, seattlepi.com, and KIRO-TV had particularly prolific reporting. The story was reported internationally.

The variable of **how much Internet discussion was linked to this event** is that nearly 3,000 comments were posted among the top three copies of the video on YouTube; there are additional comments posted on the many other copies. The Google search +Cobane +police +video retrieves over 1,400 blog posts, over 400 discussion forum posts, over 400 videos, and over 200 Facebook entries.

CASE STUDY VII: QUODOMINE, SHARIFF, 2008-10-26

On October 26, 2008, on a public sidewalk in Newark, New Jersey, professional journalist Jim Quodomine and civilians videorecord and photograph Newark Special Police Officer Brian Sharif assaulting Quodomine. This case is included for contrast (differing results for expected reasons), per Yin (2003, pp. 5, 110), based on the hypothesis that user-generated online video is significantly different from professional video journalism in its effects on accountability in police-civilian interactions in American public space. The videographer is a professional journalist, but otherwise the phenomenon is a police-civilian interaction in American public space during which police misconduct may have occurred.

It is particularly important to examine this case because it exemplifies attempted prior restraint on First Amendment protected speech by police. This is precisely the point raised in the first justification for the present research, section 1.2.: The stated problems have significant implications for the continued exercise of First Amendment rights in photographing public space, both for autonomous citizens and for professional journalists.

The scope of the present research explicitly includes police. To reiterate from section 1.3.3., historically and recently, the state has extended police powers to a broad range of persons. Thus, the term 'police' as used in this research includes any law enforcement officer, private security guard, or other person granted police powers to act for a government agency while interacting with civilians in American public space. As previously noted, present estimates are that each public police officer has approximately three counterparts in private security (Goldstein, 2007, para. 7). There has also been a trend to the privatization of policing, even of public spaces such as parks and transit stations (Mitchell, 2003, p. 1).

In this case, the government agency is the City of Newark. In a press conference following the incident, Newark Mayor Cory Booker explained that Newark Special Police Officers are trained and can be disciplined by the Newark Police Department, but have a separate command structure. According to the mayor, the Special Police are available for hire by both public agencies and private groups, and at the time of the incident, had been contracted by the church that sponsored the demonstration (Rothman, 2008, paras. 6-7). Following the broadcast of the video, Sharif was suspended by the Newark Police Department (Rothman, 2008, para. 1), evidence that he was under their authority. The City of Newark also indemnified Sharif, as is standard practice with municipal police officers, and settled Quodomine's civil rights suit (*Quodomine v. City of Newark et al.*, 2011 May 12), taking financial responsibility for Sharif's actions.

The public sidewalk where this incident occurred (Rothman, 2008, para. 4) is American public space within the scope of this research. Thus, Newark Special Police are granted police powers to act for a government agency while interacting with civilians in American public space, and are therefore police within the scope of the present research.

Research Question and Hypotheses

For Research Question 1, what is the outcome of user-generated online video on police-civilian interactions in American public space?, the outcomes of this case include: the suspension without pay of Sharif (Epstein, 2008; WCBS-TV, 2008b); dismissal of Quodomine's summons for disorderly conduct (Ryan, 2009; *Quodomine v. City of Newark et al.*, 2009 October 22, p. 12); and Quodomine's civil rights suit against Sharif and the city, which was settled before trial (*Quodomine v. City of Newark et al.*, 2011 May 12).

For this case, Hypothesis 1, that user-generated online video has the potential to improve accountability in police-civilian interactions in American public space, is evidently positive. The video documented police misconduct in a form city officials were not able to counter, with the outcome that Sharif was suspended (Epstein, 2008; DJacobs2009, 2009) and Quodomine was compensated (*Quodomine v. City of Newark et al.*, 2011 May 12). The video was available online, including copies posted by several YouTubers (RepublicanRanting, 2008; MarzuqVision, 2008; DJacobs2009, 2009).

The value of Hypothesis 2, that user-generated online video is significantly different from professional video journalism in its effects on accountability in police-civilian interactions in American public space, in this case does not appear to be evaluable without conjecture.

Hypothesis 3, that there are strong motivations for police to continue to attempt to restrict civilians' First Amendment rights to photograph police in public spaces, appears to be ambiguous in this case: Sharif lost pay while suspended (Epstein, 2008; DJacobs2009, 2009), but was reinstated (Edgar, 2011).

Documents

There are over 40 documents containing relevant data for this case, including five media clips from YouTube and broadcast TV (WCBS-TV, 2008a, 2008b; RepublicanRanting, 2008; MarzuqVision, 2008; DJacobs2009, 2009). There are over 90 court documents, but most relevant data are contained in the complaints and answers. Aside from the three CBS broadcasts and the initial AP wire story (WCBS-TV, 2008a; WCBS-TV, 2008b; RepublicanRanting, 2008; MarzuqVision, 2008; DJacobs2009, 2009; Epstein, 2008), there has been very little media coverage outside of Newark. The video was shot in high definition by a freelance cameraman on assignment for CBS, and supplemented by photographs taken by civilian bystanders. The story was broadcast at 6:33 PM and at 11:04 PM on WCBS 2 that day (WCBS-TV, 2008a); a follow-up was the lead story on the 6 PM broadcast the following day (WCBS-TV, 2008b). These broadcasts were duplicated on YouTube and remain available as of February 2, 2012 (RepublicanRanting, 2008; MarzuqVision, 2008; DJacobs2009, 2009).

Unfortunately, none of the video documents are available in the original high definition format. Sharif issued a summons to Quodomine for disorderly conduct; this should be one of the few official documents in the case, but it was never made public, even in court records. Sharif was on 'special duty' that is, hired, for the church where the march was ending; there would have been documents to that effect, both at the church and the NPD, but none have been made public. However, Newark city officials stated in a televised press conference that Sharif had been suspended without pay (Epstein, 2008; Jackson, 2008a; DJacobs2009, 2009). Numerous witnesses, including the Newark Municipal Council President, were interviewed on camera following the incident. The reporter also stated that organizers of the march had invited the press, which was corroborated by community forum posts; they wanted coverage (WCBS-TV, 2008a, b; Miss Tam-Tam, 2008).

Relevant Variables of Interest

The variable of **whether the police were aware of the camera** is positive (WCBS-TV, 2008a). The variable of **when the police became aware of the camera** is from the beginning of the incident; the camera was an Electronic News Gathering (ENG) system, a bulky shoulder-mounted rig that is highly obtrusive (WCBS-TV, 2008a). The variable of **whether there was any marked change in police behavior once they were aware of the camera** is negative; Sharif's behavior began with unlawful instructions to the videographer (WCBS-TV, 2008a). The variable of **whether police made any attempt to prevent the recording** is positive. Sharif repeatedly ordered the videographer to "put away the camera," then grabbed the camera, took it away from the videographer, and forcefully detained the videographer, preventing the videographer from recording (WCBS-TV, 2008a; Epstein, 2008).

The variable of **whether police made any attempt to acquire, confiscate, or destroy the video** is apparently negative, based on the document search and acquisition procedures. Sharif's temporary seizure of the camera in taking it away from the videographer was apparently not followed up with any police attempt to acquire, confiscate, or destroy the video itself. None of the reports or court documents make that claim, and the video was broadcast within a few hours of the incident.

The variable of **whether the police gave any unlawful instruction to the videographer regarding the video** is positive. Sharif repeatedly ordered the videographer to "put away the camera" (WCBS-TV, 2008a; Epstein, 2008) in clear violation of the First Amendment protections for newsgathering activities.

The variable of **whether police detained, cited, or arrested the videographer** is positive. Sharif detained and arrested the videographer, handcuffing him and placing him in a police vehicle, and issued him a summons for disorderly conduct pursuant to N.J.S.A.

2C:33-2. (WCBS-TV, 2008a; Epstein, 2008; *Quodomine v. City of Newark et al.*, 2009 October 22, pp. 3, 9).

The variable of **when the video was available via the Internet** is October 26, 2008; the network video was available on the WCBS-TV website, and the first copy was uploaded to YouTube less than a day later (WCBS-TV, 2008a; Epstein, 2008; RepublicanRanting, 2008).

The variable of **when the video was available via broadcast news media** is 6:33 PM and 11:04 PM, October 26, 2008, and 6:00 PM, October 27, 2008, on WCBS 2 (WCBS-TV, 2008a, 2008b; MarzuqVision, 2008; AmericanFascism, 2009; RepublicanRanting, 2008).

The variable of when images from the video were available via print news media is October 27, 2008 (Rothman, 2008).

The variable of **whether more than one camera captured the event** is positive. At least one other civilian camera captured the incident, and WCBS 2 intercut those still images with Quodomine's video footage for the broadcast videos. These user-generated images continued the visual documentation of the incident after Sharif had taken the video camera away from Quodomine (WCBS-TV, 2008a).

The variable of **whether official CCTV, dashboard, or other video captured the event** is negative, based on the document search and acquisition procedures; no official video of the incident was publicly available. The variable of **whether police initially admitted to the existence or possession of video of the event** is therefore not applicable, and the same is true for the variables of **when did police admit to possession of video of the event, whether police released official video of the event**, and **whether official video of the event was available via the Internet**.

The variable of **who was credited as the source for each medium of release of the video** was WCBS-TV for the direct links to the WCBS website, and YouTube for all other links that gave any source credit. The CBS corner logo remains visible in most of the copies online. No medium of release credited the videographer by name (Rothman, 2008; Ryan, 2008; Edgar, 2011; WCBS-TV, 2008a, 2008b; RepublicanRanting, 2008; MarzuqVision, 2008; DJacobs2009, 2009; AmericanFascism, 2009).

The variable of **whether there was any effort to restrict, remove or prosecute the release of the video** is negative, based on the document search and acquisition procedures.

The variables of the **initial official response to handling of the videographer** and the **initial official response to handling of the civilian subject** were statements from Newark Municipal Council President Mildred Crump, who was attending church services at the scene of the incident. She advocated for the videographer's release, and talked to Sharif, then spoke with the reporter:

> I did show him the picture of the, um, your cameraman, who was cuffed, and, in, in what appeared to be a chokehold. His side of the story is that he had asked him not to do it, and when he, uh, persisted, decided that he just wouldn't take it anymore. One of the reasons I said, 'we need to release your, uh, uh, cameraman,' is because you have a Constitutional right to, uh, videotape, uh, what is going on. (WCBS-TV, 2008a)

Crump also said that she would ask for an investigation.

The Newark police department declined to comment at the time, but at a press conference the following day Police Director Garry McCarthy joined Mayor Cory Booker in making an official statement of Sharif's suspension without pay, and that the city was investigating (Epstein, 2008; Rothman, 2008). Booker promised, "We will deal with this person swiftly now. We will deal with this person, if they have crossed that line, in a way that sends a message, not only to every Special Police officer, but everyone" (WCBS-TV, 2008b; DJacobs2009, 2009). In addition, the mayor stated that he was "disgusted" and "disturbed" after viewing the video of the incident, and that "People don't always follow their training. Just because you wear a badge doesn't mean you are perfect" (Epstein, 2008; WCBS-TV, 2008b; Jackson, 2008a).

The variables of the **final outcome regarding the videographer** and the **final outcome regarding the civilian subject** are the civil rights suit was settled before trial, with no admissions of guilt by any of the defendants (*Quodomine v. City of Newark et al.*, 2011 May 12).

The variables of the **sequence of official actions regarding the videographer** and the **sequence of official actions regarding the civilian subject** are complex. Newark Special Police officers, including Sharif, had been contracted by the Metropolitan Baptist Church. A call to Newark churches about the day's demonstration had been listed on community calendars; citizens and the press had been invited. Prior to the incident, a newspaper reporter had taken photos of the marchers (Miss Tam-Tam, 2008; Jackson, 2008c). Sharif was on duty when the marchers approached peacefully along Springfield Avenue, accompanied by the CBS news team. The videographer was standing on the sidewalk and the camera was recording when Sharif ordered him to put the camera away. When Quodomine ignored the unlawful instruction, Sharif forcibly arrested him, placing him in a choke hold, saying "I hate the press . . . I can do whatever I want" (WCBS-TV, 2008a; Epstein, 2008), and threatening to "break his arm" (*Quodomine v. City of Newark et al.*, 2009 October 22, p. 3). Sharif also threatened to arrest CBS reporter Christine Sloan (WCBS-TV, 2008a; Epstein, 2008). One other police officer is visible in the video, and does not appear to be attempting to prevent any of Sharif's actions (WCBS-TV, 2008a). Following the incident, Sharif issued the videographer a summons for disorderly conduct pursuant to N.J.S.A. 2C:33-2 (*Quodomine v. City of Newark et al.*, 2009 October 22, p. 9).

The Office of the Newark Municipal Prosecutor was provided with copies of the video that clearly showed Sharif's actions, and which contradicted Sharif's statements in his incident report. However, the prosecutor stated to the videographer's attorney that the charges would only be dismissed if the videographer agreed to release the city from any and all potential civil liability (*Quodomine v. City of Newark et al.*, 2009 October 22, pp. 10-11).

The disorderly charges were a Municipal Court matter. However, at some of the proceedings and in conferences with the Municipal Judge assigned to the case, corporate counsel for the city Ann Periera attended and participated. Sharif's personal attorney also attended and participated in these discussions. Together, the prosecutor, corporate counsel, and Sharif's attorney offered that the charges would be dropped if the videographer either agreed to release the city from civil liability, or stipulated that Sharif had probable cause for the arrest. Either of these options would have prevented the videographer from recovering civil damages; his attorney refused the offers. Corporate counsel then indicated that a trial would be necessary. According to the videographer's attorney, neither Newark corporate counsel nor Sharif's attorney had legal standing in the case, and their participation was ethically questionable. These actions were cited when corporate counsel and the prosecutor were later named as defendants in the videographer's lawsuit. On the scheduled trial date of February 4, 2009, the municipal judge advised that there was a potential conflict of interest, as she was also a city employee, so the case would be transferred to a 'conflicts judge' in Newark (*Quodomine v. City of Newark et al.*, 2009 October 22, pp. 11-12; ACLU-NJ, 2010, p. 54).

The Essex County municipal court judge thereafter received a letter from the videographer's attorney, documenting these events, pointing out the corporate counsel's violations of professional ethics, and asking that the case be moved out of Newark to a court that had no connection to the city. The case was thereafter transferred to the Superior Court, Law Division, Essex County, and prosecution was taken over by the Essex County Prosecutor. That prosecutor determined that the charges were unsubstantiated, and dismissed them (*Quodomine v. City of Newark et al.*, 2009 October 22, p. 12; ACLU-NJ, 2010, p. 54).

Official responses to each of the points in the videographer's civil suit were categorical denials without significant new data for the variables of interest (*Quodomine v. City of Newark et al.*, all dates). After the customary legal maneuvers and following the completion of discovery (PACER, 2011), the defendants settled the civil rights suit for money only, with no admissions of guilt (*Quodomine v. City of Newark et al.*, 2011 May 12). The Newark Municipal Council voted to authorize certification of funds for the settlement on August 3, 2011 (City of Newark, 2011), nearly three years after the incident.

The variable of the **compensation of the videographer** for the recording of the video is not public, as it is a matter of his contract terms with WCBS-TV.

The variable of the **affiliation of the videographer** is professional freelance journalist under contract to WCBS-TV, the flagship station of the CBS network (WCBS-TV, 2008a, b; *Quodomine v. City of Newark et al.*, 2009 October 22, p. 3).

The variable of **police misconduct** is evidently positive. The sequence of police actions recorded in the video evidently constitutes misconduct, including excessive use of force, false arrest, and violation of First Amendment protections of free speech, among others (*Quodomine v. City of Newark et al.*, 2009 October 22; ACLU-NJ, 2010; WCBS-TV, 2008a; DJacobs2009, 2009).

The variable of the **initial response of the law enforcement agency regarding the officers involved** is that Sharif was suspended without pay pending investigation (Epstein, 2008; Jackson, 2008a; Rothman, 2008; WCBS-TV, 2008b).

The variable of the **sequence of official actions regarding the officers involved** was that after an unpublicized period of suspension, Sharif was returned to duty (Edgar, 2011). The duration of the suspension was not evident within the results of the document search and acquisition procedures.

The variable of the **final outcome regarding the officers involved** was that Sharif returned to duty, and as of July 2011 remains a Newark Special Police Officer, assigned to government schools (Edgar, 2011).

For the variable of **final outcome of the law enforcement agency regarding policy changes, if any**: the settlement included no admission of guilt from either the department or Sharif, and no policy changes have been evident (*Quodomine v. City of Newark et al.*, 2011 May 12). Due to this and other evidence of corruption in the Newark Police Department, the ACLU of New Jersey has petitioned the US Department of Justice to investigate the department (ACLU-NJ, 2010).

The variable of the **affiliation of the civilian subject** is that the videographer and civilian subject was a professional journalist, on assignment for a television network news team (Epstein, 2008). CBS Stations Group vice president Michael Nelson publicly stated, "WCBS-TV stands behind the conduct of Jim Quodomine, who clearly did nothing wrong" (Epstein, 2008).

The variable of the **compensation (if any) of the subject** is a settlement in the amount of $121,644.48. This amount was not initially made public through court documents, but became public when the city council of Newark voted to authorize the required funds (City of Newark, 2011).

The variable of **what third parties (if any) involved themselves in the legal case** is that the ACLU of New Jersey was involved in the civil suit, and highlighted the case on its website, including links to one of the videos and to the first complaint of the civil suit (ACLU-NJ, 2011; MarzuqVision, 2008; *Quodomine v. City of Newark et al.*, 2009 October 22).

The variable of ***cui bono?*, or who benefits?**, in this case is the videographer and, presumably, his attorney, at least monetarily (*Quodomine v. City of Newark et al.*, 2011 May 12; City of Newark, 2011). Any community benefits are not evident, as Sharif was reinstated, none of the defendants admitted or apologized for any wrongdoing, and no policies have been changed.

The variable of **how many times the video was viewed online** is that the primary YouTube video has been viewed nearly 29,000 times (RepublicanRanting, 2008), but the video is periodically duplicated and uploaded again as new viewers discover it, so the combined total is in excess of 78,000 views. WCBS-TV website viewer data were not made public, and are no longer available. WCBS-TV's own site no longer carries the video nor any other reference to Quodomine or Sharif; however, the YouTube videos remain available (RepublicanRanting, 2008; DJacobs2009, 2009; MarzuqVision, 2008; AmericanFascism, 2009).

The variable of **how much press coverage the event received** was in single digits in number of reports. There was very little traditional mass media coverage of this story, aside from the television broadcasts of the station whose videographer was arrested. It was the lead story on the 6 PM WCBS-TV news broadcast the day after the incident, and featured a news conference with the mayor of Newark that showed flagged microphones from at least four other television stations (DJacobs2009, 2009). The news conference also included Police Director Garry McCarthy and Essex County Prosecutor Paula Dow, and had evidently been called to announce the apprehension of suspects in the previous Friday's string of shootings (Jackson, 2008a, 2008b). The original incident story was picked up by the AP wire service (Epstein, 2008). There were two articles, one in the *Newark Star-Ledger* and the other in the *New Jersey Herald*, when the civil suit was filed (Ryan, 2009; Howell, 2009); neither story was picked up by the wire services. Based on the document search and acquisition procedures, there were no follow-up reports when the disorderly conduct charges were dropped or when the civil rights case was settled. The ACLU of New Jersey has not updated the case summary on its website (ACLU-NJ, 2011).

The variable of **how much Internet discussion was linked to this event** is, 18 blog entries and 40 discussion forums as of February 1, 2012 have mentioned this case (Google +Quodomine +Sharif +Newark). This story only gained prominence outside the New York/Newark area after copies of the broadcast segments had been uploaded to YouTube; then links and comments proliferated rapidly. A majority of the online commentary is highly critical of the police; a few useful comments provide additional context, including links to news and photographs of the more peaceful beginnings of the march prior to the incident (Miss Tam-Tam, 2008; Jackson, 2008c; MarzuqVision, 2008; AmericanFascism, 2009; RepublicanRanting, 2008; DJacobs2009, 2009; Edgar, 2011).

CASE STUDY VIII: HAKEL, MCCARREN, ASHTON ET AL., 2005-04-15

On April 15, 2005, on a public road in Prince George's County, Maryland, professional journalist Pete Hakel videorecorded county police carrying out a felony stop of television reporter Andrea McCarren while she was gathering information for a story of public interest. This case is included for contrast (differing results for expected reasons), per Yin (2003, pp. 5, 110), as both Hakel and McCarren are professional journalists. The phenomenon is a police-civilian interaction in American public space during which police misconduct may have occurred, video of which interaction was available online.

Research Question and Hypotheses

For Research Question 1, what is the outcome of user-generated online video on police-civilian interactions in American public space?, the outcomes of this case include: PGC officials delayed legal proceedings for four years (*McCarren v. Prince George's County et al.*, all dates); a citizen's advisory panel recommended that two officers be disciplined for not having dashboard cameras running (Segraves, 2007; McCarren, 2007); the county said two officers were disciplined, but refused to release details (Segraves, 2007); a jury awarded the police-injured TV reporter only actual damages for medical expenses, plus costs (*McCarren v. Prince George's County et al.*, 2009); the police corporal who initiated the incident was promoted to Sergeant and transferred to Internal Affairs (*Springer v. Prince George's County et al.*, 2009).

For this case, Hypothesis 1, that user-generated online video has the potential to improve accountability in police-civilian interactions in American public space, is evidently ambiguous. Two officers were reportedly disciplined (Segraves, 2007; McCarren, 2007), and the reporter was compensated (*McCarren v. Prince George's County et al.*, 2009). However, the police corporal who initiated the incident was promoted and transferred to Internal Affairs (*Springer v. Prince George's County et al.*, 2009).

The value of Hypothesis 2, that user-generated online video is significantly different from professional video journalism in its effects on accountability in police-civilian interactions in American public space, in this case does not appear to be evaluable without conjecture.

Hypothesis 3, that there are strong motivations for police to continue to attempt to restrict civilians' First Amendment rights to photograph police in public spaces, does not appear to be evaluable without conjecture. Based on the document search and acquisition procedures, there are no documents recording definite sanctions for the police.

Documents

The case data exceed 40 files, including two media clips from YouTube and broadcast TV. There were over 40 news wire articles from LexisNexis Academic News alone. There were 138 court documents, the majority of them legal maneuvering with little unique data for the variables of interest. Documents from the October 2006 mediation (on which the County later reneged) are not publicly available. McCarren has written on this case in the journalism literature. The video was shot by Pete Hakel, videographer for WJLA-TV, during a felony traffic stop falsely called in as an "officer in trouble" distress call by Corporal Danon Ashton, who the journalists were following in a corruption investigation. Hakel continued to record, but was not free to direct the camera to capture the most important action. The actual use of force that injured McCarren is not on the video. The video was later broadcast as part of the story presented by WJLA. Those broadcast segments were available for a time on WJLA's YouTube channel, but have since been removed and are no longer available on the Internet. There were no charges filed against McCarren or Hakel, so there are no charging documents. There were also no dashboard videos from any of the nine cruisers, despite the fact that Prince George's County Police were required by the U.S. Justice Department to have dashboard cameras rolling for all felony stops. Police never complied with requests for 911 tapes or cell phone records. Medical records introduced in court documented that McCarren had surgery to repair a torn rotator cuff, torn labrum and detached biceps tendon.

Relevant Variables of Interest

The variable of **whether the police were aware of the camera** is positive. Police were recorded issuing specific instructions about the camera (WJLA, 2005).

The variable of **when the police became aware of the camera** is, it is evident from the video that at least some of the police participating in the felony stop were not initially aware of the videographer in the back seat of the SUV, or that he had a camera; when they became aware of him, they ordered him out of the vehicle, and ordered him to "drop the camera" (WJLA, 2005, 1:53).

The variable of **whether there was any marked change in police behavior once they were aware of the camera** is positive. For the police participating in the felony stop, their marked change in behavior once they were aware of the camera was to tell the videographer to put it down (WJLA, 2005, 1:53). After the journalists were released and the reporter began asking questions, the lead officer waved off the camera and stated, "We're not taping, I'm not giving an interview" (WJLA, 2005, 2:21).

The variable of **whether police made any attempt to prevent the recording** is positive. It is evident that the police conducting the felony stop ordered the videographer to put the camera down (WJLA, 2005, 1:53). One of the police then placed the video camera in the back seat of the SUV, where police would not be visible to its lens (McCarren, 2007).

The variable of **whether police made any attempt to acquire, confiscate, or destroy the video** is negative. Aside from placing the camera back in the SUV, the police did not evidently tamper with the camera, and did not appear to order either of the journalists to surrender or to destroy the recording (WJLA, 2005; McCarren, 2007).

The variable of **whether the police gave any unlawful instruction to the videographer regarding the video** is negative. The police order to the videographer to "drop the camera" (WJLA, 2005, 1:53) was a lawful instruction in the context of a felony stop, which police "use when they feel there is a threat…the objective is to get people out of the vehicle with their hands visible and to gain control of them," according to Percy Alston, president of Fraternal Order of Police Lodge 89 (Klein, 2005).

The variable of **whether police detained, cited, or arrested the videographer** is positive. The videographer was detained until his identity was determined to the satisfaction of the police conducting the felony stop. He and the reporter were thereafter released, and were not cited (WJLA, 2005; McCarren, 2007).

The variable of **when the video was available via the Internet** is May 13, 2005, the day of the first television broadcast (WJLA, 2005).

The variable of **when the video was available via broadcast news media** is May 13, 2005, during WJLA's 5 PM television news broadcast (WJLA, 2005; McCarren, 2007).

The variable of **when images from the video were available via print news media** is May 13, 2005, the day of the first television broadcast; the images appeared on the front page of the *Washington Post*'s Metro section (McCarren, 2007, p. 4; Castaneda, 2005).

The variable of **whether more than one camera captured the event** is apparently negative. Although the police were required to have dashboard-mounted video cameras operating in their vehicles, no recordings from those cameras were ever produced (McCarren, 2007; ABC-7, 2005; Klein, 2005; Associated Press, 2005; Segraves, 2007, 2009; Pavsner, 2007; Castaneda, 2007, 2009; Burns, 2009; Balko, 2011).

The variable of **whether official CCTV, dashboard, or other video captured the event** is apparently negative. Under a 2004 memorandum of agreement with the US Department of Justice, stemming from federal investigations of systemic misconduct, PGC police were required to have dashboard-mounted video cameras operating in their vehicles during felony traffic stops. No recordings of this incident by those cameras were ever produced; the official statement was that all seven cameras experienced technical failure at the same time (McCarren, 2007; ABC-7, 2005; Klein, 2005; Associated Press, 2005; Segraves, 2007, 2009; Pavsner, 2007; Castaneda, 2007, 2009; Burns, 2009; Balko, 2011). For the same reasons, the variables of **whether police initially admitted to the existence or possession of video of the event, when did police admit to possession of video of the event, whether police released official video of the event**, and **whether official video of the event was available via the Internet** are all not applicable.

The variable of **who was credited as the source for each medium of release of the video** is that the ABC 7 logo was superimposed in the lower right corner of the video (WJLA, 2005). The videographer was also named in the video and in the majority of the news reports about the incident.

The variable of **whether there was any effort to restrict, remove, or prosecute the release of the video** is evidently negative. The video was available via broadcast, Internet, and print within a 24-hour period thirty days after the incident. Based on the document search and acquisition procedures, there are no public documents recording any effort to prevent any of those releases.

The variable of the **initial official response to handling of the videographer** is that county officials refused to release any information about the incident (McCarren, 2007; ABC-7, 2005; Klein, 2005; Associated Press, 2005).

The variable of the **final outcome regarding the videographer** is that he was not injured and was not a party to the legal actions (McCarren, 2007; *McCarren v. Prince George's County et al.*, 2007).

The variable of the **sequence of official actions regarding the videographer** is almost entirely summarized by the broadcast video (WJLA, 2005). The videographer was a passenger in the middle of the back seat of the SUV driven by the reporter, and was recording at the time of the felony stop. After the reporter exited the vehicle at police orders and was handcuffed, the police ordered the videographer out of the vehicle. When the police observed that the videographer had a camera, they ordered him to "drop the camera" (WJLA, 2005, 1:53; McCarren, 2007). The videographer placed the camera, which continued to record, on the ground and pointed in the direction of the police. As police frisked the videographer, one of them noticed the camera, and gestured to another officer to remove it. That officer tossed it into the back seat of the SUV. Police released the videographer after they frisked him. After identifying the reporter, police told both the videographer and the reporter to leave so the police could get traffic moving. The reporter began asking the police questions, and the videographer began recording again (McCarren, 2007). The lead officer waved off the camera and stated, "We're not taping, I'm not giving an interview" (WJLA, 2005, 2:21). The reporter asked for a public information officer to come to the scene, as legally required when a member of the press was involved; police informed them that no one was available. The videographer drove the reporter's SUV, reportedly because the reporter was shaken and in pain from her shoulder injury.

As they drove away from the scene, the videographer saw Ashton and Brown smiling at them (McCarren, 2007). There is no documentation of any further official actions regarding the videographer.

The variable of the **compensation of the videographer** is not public, as it is a matter of his employment contract with WJLA.

The variable of the **affiliation of the videographer** is that he was a professional journalist employed by the Washington affiliate of the ABC network, WJLA ABC-7 (WJLA, 2005; ABC-7, 2005; McCarren, 2007).

The variable of **police misconduct** is positive. It was the finding of the jury that police used excessive force (*McCarren v. Prince George's County et al.*, 2009 January 28; Segraves, 2009; Castaneda, 2009; Burns, 2009; Balko, 2011).

The variable of the **initial response of the law enforcement agency regarding the officers involved** is that county officials refused to release any information about the incident for a month, denying WJLA's FOIA requests for 911 tapes and cell phone records of the police involved, plus cell phone records for Brown, Ashton, Johnson and Keary (McCarren, 2007; ABC-7, 2005; Klein, 2005; Associated Press, 2005). After the expiration of the legal 30-day deadline for complying with FOIA requests under the Maryland Access to Records statute, WJLA filed an internal affairs complaint (Castaneda, 2005; National Press Club, 2005; Klein, 2005, Associated Press, 2005; Brandus, 2005).

The variable of the **sequence of official actions regarding the officers involved** is complex. On May 13, police spokeswoman Barbara Hamm told the *Washington Post* that "We have not had the opportunity yet to investigate, but on the face of it, it appears that our officers followed proper procedure" (Castaneda, 2005). On May 19, Hamm said that there was no tape of the incident because none of the cameras in the seven PGC police vehicles was both functional and turned on, and that "we are in the process of investigating why" (Klein & Wiggins, 2005). In July, county officials informed WJLA's attorney that the official videotapes were missing (Washingtonian, 2005).

PGC Executive Jack B. Johnson, one subject of the reporter's investigations, appeared on radio and television broadcasts saying he believed his officers "acted appropriately" and that "the use of force was reasonable" (McCarren, 2007). Police Chief Melvin High, a Johnson appointee, repeatedly stated that he would "get to the bottom of this" (Klein & Wiggins, 2005) and "if our people didn't do what they were supposed to do ... they're held accountable," and both Johnson and High promised a "thorough investigation" (McCarren, 2007).

PGC internal police investigations go before the Citizen Complaint Oversight Panel (CCOP), which makes a recommendation to the county. For two of the officers, the panel recommended disciplinary action for not having their dashboard cameras running. Nine months after the felony stop, the county said two officers were disciplined (Segraves, 2007). However, the details of the disciplinary actions, if any, do not appear in public documents (McCarren, 2007). Although later editions of the citizen panel's mandated Annual Report to the Public include selected case summaries, there is no publicly available summary for this investigation.

During the civil case, the court ordered the defendants to "provide all documents that refer or relate to any problems with the video camera in Officer Jermaine Allen's cruiser on or around April 15, 2005 including but not limited to the documents Officer Allen referred to in his deposition" (Maryland Judiciary, 2011). The court also ordered a redacted copy of the internal affairs report be filed in the case jacket under seal (Maryland Judiciary, 2011), so although it was introduced in court, the internal affairs report is not a publicly available document.

The variable of the **final outcome regarding the officers involved** is in large part unknown, because documents that might reveal those outcomes have either not been publicly released, or were sealed by the court. However, at least one later lawsuit naming Ashton describes him as of December, 2008 as Sgt. Danon Ashton of Internal Affairs (*Springer v. Prince George's County et al.*, 2009).

For the variable of **final outcome of the law enforcement agency regarding policy changes, if any,** is that during 2005, the CCOP asked the Chief of Police for changes in the department's use of force policy and "the number of unused and inoperable video cameras in police cruisers" (OMB, 2005, p. 46). The issue of lack of functioning mobile video systems during traffic stops improved enough through 2008 and 2009 that it was removed as a recurring issue from the 2009 report. The issue became a problem again, and was reported in 2010. The CCOP also identified "an increasing number of investigations involving the failure to have adequate video monitoring equipment in police cruisers" (CCOP, 2010, p. 2).

The variable of the **affiliation of the civilian subject** is professional journalist. She was an investigative reporter for the Washington affiliate station of the ABC television network, WJLA ABC-7 (WJLA, 2005; ABC-7; McCarren, 2005).

The variable of **the initial official response to handling of the civilian subject** is that county officials refused to release any information about the incident for a month, denying WJLA's FOIA requests for 911 tapes and cell phone records of the police involved, plus cell phone records for Brown, Ashton, Johnson and Keary (McCarren, 2007; ABC-7, 2005; Klein, 2005; Associated Press, 2005). After the expiration of the legal 30-day deadline for complying with FOIA requests, WJLA filed an internal affairs complaint (Castaneda, 2005; National Press Club, 2005; Klein, 2005, Associated Press, 2005).

The variable of the **sequence of official actions regarding the civilian subject** is complex. Corporal Danon Ashton was alleged to be acting as personal chauffeur for County Chief Administrator Jacqueline Brown, in violation of official policy on use of county resources (Castaneda, 2005). This information had been provided to the reporter, who with the videographer followed Ashton beginning around 8:20 AM on April 15, 2005 in a residential area in Bowie, MD. Ashton broke contact by pulling into a private drive, then picked up Brown out of the reporter's sight, circled around, and pulled up behind the reporter's vehicle where she had stopped to read a map (McCarren, 2007). McCarren was driving a car registered in her name, is a well-known television reporter in the area, and had been engaged in correspondence with county officials (McCarren, 2007).

Ashton chose to call the communications supervisor directly using his cell phone, bypassing dispatch and the 911 system's mandatory recording. Police sources later reported that the call went out as "an officer in trouble" but with no description of the suspect. At WJLA, two assignment editors listening to scanners heard police transmissions about "suspects with a video camera" (McCarren, 2007).

Seven county police cruisers and two from Cheverly surrounded and forced the reporter's vehicle off Landover Road near US Route 50, and shut down traffic in both directions (Castaneda, 2005). As many as twelve police kept guns pointed at the reporter and videographer in the course of the stop (McCarren, 2007). Police instructed the reporter to exit the vehicle, keep her hands up, and back toward the police. The videographer, in the SUV's back seat, was recording. The reporter backed out of the camera's view (WJLA, 2005), at which point one of the police pulled her right arm behind her back with enough force to tear her rotator cuff, labrum, and to detach a biceps tendon, all of which later required surgery to repair (Segraves, 2009). The reporter is five feet four inches and weighs about 110 pounds (Castaneda, 2005). Police handcuffed the reporter, pushed her over the hood of a cruiser, and frisked her. When police noticed the videographer, they also directed him to exit the vehicle; he placed the camera on the ground, pointed toward the police, and was handcuffed and frisked much more gently than the reporter had been (McCarren, 2007). The camera continued to run, recording video until police tossed it into the back seat, after which the audio continued to record police conversations. The audio recording documented that "most of [the police] felt confused and angry by their order to chase down what turned out to be a television news-crew pursuing a story" (McCarren, 2007).

Police emptied the reporter's purse and examined its contents, including her press credentials for the White House, Capitol, and Pentagon. Once the reporter and videographer were identified, police uncuffed them and told them to leave. Neither the reporter nor the videographer was charged. Police refused to bring a public information officer to the scene when the reporter requested one (McCarren, 2007).

WJLA immediately asked officials about the incident, and submitted FOIA requests. Several officials promised an investigation (Castaneda, 2005; Klein & Wiggins, 2005; McCarren, 2007), but the county did not comply with the station's requests for information. After county officials refused to release any information for a month, WJLA broadcast the story (WJLA, 2005).

County officials began to portray the reporter as a potential terrorist. County spokesman Jim Keary said, "She might be 5-4, but threats come in all sizes" and compared the reporter to the pilot who had recently flown his Cessna into restricted air space near the White House (Klein, 2005). Public Safety Director Vernon Herron told the *Washington Post* that government officials are "threatened and assaulted every day, some even killed in the performance of their duties" but admitted that Brown had never been threatened (McCarren, 2007). In July, a PGC police source said McCarren should stop whining because "she wasn't shot" (Washingtonian, 2005). Police cars drove slowly by the reporter's house or parked there in what the reporter described as "a not-so-veiled threat" (McCarren, 2007). Police sources also refused to cooperate with the reporter's colleagues, who in at least one instance blamed her (McCarren, 2007).

The reporter and her legal team prepared a lawsuit, but just prior to filing the county requested mediation. The reporter was advised to agree, which she did, and on October 3, 2006 the reporter, her attorney, and county attorneys, mediated by a retired federal judge, settled on a payment and an apology. Both sides signed the agreement, and the check and the letter of apology were to be delivered to the reporter by mid-November (Castaneda, 2007; McCarren, 2007). County attorneys delayed for six months, then refused to honor the agreement; the payment was acceptable, but the county refused to make an apology and wanted the entire agreement made confidential (Castaneda, 2007).

McCarren then sued (Castaneda, 2007; *McCarren v. Prince George's County et al.*, 2007). Official responses were to delay and obstruct legal proceedings, as evidenced by the court docket, and to make the process damaging and embarrassing for the reporter, including attempts to publicly release all her telephone records (which would be severely damaging to an investigative journalist with confidential sources) and her medical records including photographs (Maryland Judiciary, 2011). At the same time, officials attempted to prevent the introduction of their own procedural manuals and to bar statements made by police (Maryland Judiciary, 2011; *McCarren v. Prince George's County et al.*, all dates).

Following the jury award in January 2009, Keary told WTOP that the verdict was a vindication for the county, and that "Sadly, Ms. McCarren was trying to grandstand and grab headlines by accusing the county of interfering with her pursuit of a story. The jury solidly said, 'no,' found the stop was proper and did not violate her rights" (Segraves, 2009). The reporter's attorney rebutted, "The verdict was anything but a vindication. The jury found that the officers who stopped Andrea used excessive force, injured her, and violated her constitutional rights. It's a sad day when the county considers such a stinging rebuke a "vindication"" (Segraves, 2009).

The variable of the **compensation (if any) of the subject** is $5,000 for medical expenses, plus court costs to the defendants (*McCarren v. Prince George's County et al.*, 2009 January 28; Segraves, 2009; Castaneda, 2009; Burns, 2009; Balko, 2011).

The variable of the **final outcome regarding the civilian subjects** was a jury verdict that police had used excessive force, but that police had not violated the reporter's civil rights under the First Amendment (*McCarren v. Prince George's County et al.*, 2009 January 28; Segraves, 2009; Castaneda, 2009; Burns, 2009; Balko, 2011).

The variable of **what third parties (if any) involved themselves in the legal case** is none. The executive director of the Reporters Committee for Freedom of the Press (RCFP) made statements about this case, but the RCFP did not file an *amicus* brief or otherwise officially participate (WJLA, 2005).

The variable of ***cui bono?*, or who benefits?**, in this case is apparently the county police and officials, who were not made fully accountable for their actions during and following this incident. However, a number of those officials were later indicted following a federal investigation, and Jack B. Johnson pled guilty to bribery, extortion, conspiracy, and witness and evidence tampering (Castaneda, 2011).

The variable of **how many times the video was viewed online** is not available, as WJLA did not make tracking information available for its own website, and WJLA has also taken down its YouTube channel (WJLA, 2005).

The variable of **how much press coverage the event received** is in the low double digits, and initially included stories on local television, a radio station, the Associated Press wire service, and in a national newspaper (ABC-7, 2005; Klein, 2005; Associated Press, 2005; Castaneda, 2005; Brandus, 2005). When the lawsuit was filed in 2007, the coverage expanded to include two press release wire services, a second national newspaper, and an industry magazine (Pavsner, 2007; Eggerton, 2007). Since the court ruling, a number of editorials and journals have cited the case (Burns, 2009; Balko, 2011).

The variable of **how much Internet discussion was linked to this event** is in the thousands of posts, and at least hundreds of threads. Websites, forums and blogs on police, First Amendment issues, and journalism are significant sources, but a broad range of discussions cite this event. Nearly 4,000 blog entries and over 200 discussion forums to date have mentioned Andrea McCarren's interaction with police (Google +Andrea +McCarren +police).

CASE STUDY IX: GLIK, CUNNIFFE ET AL., 2007-10-01

At approximately 5:30 PM on October 1, 2007, civilian Simon Glik stood on a public sidewalk on Tremont Street and held his cellular phone in plain view while using it to videorecord Boston Police Sergeant John Cunniffe, Officer Peter Savalis and Officer Jerome Hall-Brewster making an arrest of a 16-year-old alleged drug offender. This case is included for direct replication (similar results for expected reasons), per Yin (2003, pp. 5, 110). The phenomenon is a police-civilian interaction in American public space during which police misconduct may have occurred, which was videorecorded by a civilian, and the videorecording was distributed online.

Research Question and Hypotheses

For Research Question 1, what is the outcome of user-generated online video on police-civilian interactions in American public space?, the outcomes of this case include: criminal charges of wiretapping against Glik were dismissed (*Massachusetts v. Glik*, 2008); the U.S. District Court for Massachusetts has ruled that the First Amendment protection of Glik's conduct is clearly established (*Glik v. Cunniffe et al.*, 2010f, p. 7); the U.S. Court of Appeals for the First Circuit has ruled that police do not have qualified immunity from being sued for violating Glik's First and Fourth Amendment rights (*Glik v. Cunniffe et al.*, 2011e); a training video based on Glik's actions is now mandatory Boston PD viewing (FOX 25, 2010); two of the three police were departmentally disciplined and the city settled Glik's civil rights suit for $170,000 (Ott, 2012; *Glik v. Cunniffe et al.*, all dates).

For this case, Hypothesis 1, that user-generated online video has the potential to improve accountability in police-civilian interactions in American public space, is evidently positive; police were departmentally disciplined. The remaining video that Glik (2007) recorded documents a sequence of events significantly different from police statements, and the continued presence of that video on the ACLU website and elsewhere on the Internet enables the public to view that evidence without other mediation.

The value of Hypothesis 2, that user-generated online video is significantly different from professional video journalism in its effects on accountability in police-civilian interactions in American public space, in this case does not appear to be evaluable without conjecture.

Hypothesis 3, that there are strong motivations for police to continue to attempt to restrict civilians' First Amendment rights to photograph police in public spaces, for this case appears to be supported; police were departmentally disciplined. In addition, police actions in seeking to prosecute the videographer initiated the *Streisand Effect* (Masnick, 2005), drawing more attention to an event that the police sought to conceal.

Documents

The case data presently exceed 130 files, including 8 media clips from YouTube and other websites. One partial video was recovered from Glik's phone, and is now available on the ACLU website and has been duplicated on YouTube (Glik, 2007). There are presently over 40 court documents, including the original criminal case (*Massachusetts v. Glik*, 2008), and the civil suit (*Glik v. Cunniffe et al.*, all dates). Most of the documents are legal maneuvering; the complaints, answers, and rulings contain the most relevant data. There are a continuously growing number of news, opinion, and law review articles that cite this case. It is also prominent in a number of blogs, forums, and websites dealing with civil rights, technology, photography, and police issues.

Relevant Variables of Interest

The variable of **whether the police were aware of the camera** is positive. The appeals court noted that, "After placing the suspect in handcuffs, one of the officers turned to Glik and said, "I think you have taken enough pictures."'" (*Glik v. Cunniffe et al.*, 2011e, pp. 3, 22). The variable of **when the police became aware of the camera** is during the incident (*Glik v. Cunniffe et al.*, 2011e, pp. 3, 22; *Massachusetts v. Glik*, 2008). The variable of **whether there was any marked change in police behavior once they were aware of the camera** is negative (*Glik v. Cunniffe et al.*, 2011e, p. 3; *Massachusetts v. Glik*, 2008). The variable of **whether police made any attempt to prevent the recording** is negative (*Glik v. Cunniffe et al.*, 2011e, p. 3; *Massachusetts v. Glik*, 2008). The variable of **whether police made any attempt to acquire, confiscate, or destroy the video** is positive. Once police established that video had been recorded, they arrested the videographer and confiscated the camera (*Glik v. Cunniffe et al.*, 2011e, p. 3), and while the cell phone was in their possession, police deleted some of the videos (*Glik v. Cunniffe et al.*, 2010a, p. 8). The variable of **whether the police gave any unlawful instruction to the videographer regarding the video** is positive (*Glik v. Cunniffe et al.*, 2011e, p. 3). The variable of **whether police detained, cited, or arrested the videographer** is positive (*Massachusetts v. Glik*, 2008; *Glik v. Cunniffe et al.*, 2011e, p. 3).

The variable of **when the video was available via the Internet** is February 1, 2010, when the videographer's civil rights suit was filed and the ACLU put the video on their website (ACLU, 2010; Glik, 2007). The variable of **when the video was available via broadcast news media** is that WBZ 1030 News Radio's Carl Stevens discussed the case with attorney Howard Friedman the day the videographer's civil rights suit was filed, February 1, 2010 (Stevens, 2010); however, there is apparently no evidence that the video was broadcast on television until a local station ran a story on the police training video on December 22, 2010 (FOX 25, 2010).

The variable of **when images from the video were available via print news media** is apparently no earlier than February 1, 2010, when the ACLU, representing the videographer, made the video available online. However, based on the document search and acquisition procedures, it is not evident that images from the video have ever appeared in print.

The variable of **whether more than one camera captured the event** is negative. Based on the document search and acquisition procedures, only the one camera (Glik, 2007) captured the incident.

The variable of **whether official CCTV, dashboard, or other video captured the event** is negative, based on the document search and acquisition procedures. Neither the police nor Glik's court filings indicate the existence of official video of the incident (*Massachusetts v. Glik*, 2008; *Glik v. Cunniffe et al.*, all dates). For the same reasons, the variables of **whether police initially admitted to the existence or possession of video of the event**, of **when did police admit to possession of video of the event**, of **whether police released official video of the event**, and of **whether official video of the event was available via the Internet** are all not applicable.

The variable of **who was credited as the source for each medium of release of the video** is that the video has variously been credited to Glik, YouTube, and *Greater Boston* (Glik, 2007; ACLUMASS1, 2011; Waterman, 2011).

The variable of **whether there was any effort to restrict, remove or prosecute the release of the video** is positive; the police confiscated the cell phone, attempted to delete the video, and did not return the cell phone to the videographer until after the municipal court had dismissed the charges four months later (*Glik v. Cunniffe et al.*, 2010a, p. 8; *Massachusetts v. Glik*, 2008).

The variable of the **initial official response to handling of the videographer** is that police arrested the videographer, confiscated his cell phone and a flash drive, and did not release him until his wife posted a fee (*Glik v. Cunniffe et al.*, 2010a; *Massachusetts v. Glik*, 2008).

The variable of the **final outcome regarding the videographer** is that the city agreed to pay $170,000 to settle the civil rights suit, and the videographer agreed to withdraw his appeal to the civilian review board (Ott, 2012). The videographer had earlier stated that some of his goals for the case had already been achieved by the court rulings:

> One of the major goals has been accomplished by this monumental decision of the First Circuit. And it's not that I just have a right, it's everybody now who has a right and that right has been prescribed on such a level that it's impossible for the police to misinterpret (Waterman, 2011).

The variable of the **sequence of official actions regarding the videographer** is complex. Around 5:30 PM on October 1, 2007, Simon Glik stood on a public sidewalk "in the Boston Common, the oldest city park in the United States and the apotheosis of a public forum" (*Glik v. Cunniffe et al.*, 2011e, p. 14), and held his cellular phone in plain view while using it to videorecord Boston Police Sergeant John Cunniffe, Officer Peter Savalis and Officer Jerome Hall-Brewster making an arrest of a 16-year-old alleged drug offender (Glik, 2007; *Glik v. Cunniffe et al.*, 2010a, pp. 3-4; Frank, 2008; FOX 25, 2010). The officers noted that the videographer was recording them, and arrested him (*Massachusetts v. Glik*, 2008, p. 1).

Police transported the videographer to the South Boston station. One of the officers asked the videographer "if he would still be a lawyer after being charged with a felony" (*Glik v. Cunniffe et al.*, 2010a, p. 4). Police confiscated the videographer's cell phone and a computer flash drive as evidence. Despite the videographer's statement that the drive contained important computer files, the booking officer claimed that the flash drive looked like a microphone (*Glik v. Cunniffe et al.*, 2011e, p. 3; *Glik v. Cunniffe et al.*, 2010a, p. 4). The videographer was held at the police station until his wife posted a fee; his personal effects were returned, except for the flash drive and the cell phone (*Glik v. Cunniffe et al.*, 2010a, p. 4).

The videographer was initially charged with unlawful wiretap, aiding the escape of a prisoner, and disturbing the peace; the prosecutor dropped the aiding escape charge almost immediately, but chose to prosecute the wiretap and disturbing the peace charges (*Massachusetts v. Glik*, 2008, pp. 1-2). The complaint and police report did not indicate that police reviewed the contents of the videographer's cell phone to discern whether a recording was made; there was no testimony, physical evidence, or recording ever submitted (*Massachusetts v. Glik*, 2008, p. 2). The police report alleged that the videographer's actions distracted them during the drug arrest (*Massachusetts v. Glik*, 2008, p. 3).

On January 31, 2008, Mark H. Summerville, Associate Justice, Boston Municipal Court, dismissed the wiretap charge, ruling: "It seems clear that the officers were unhappy they were being recorded during an arrest. But their discomfort does not make a lawful exercise of a First Amendment right a crime" (*Massachusetts v. Glik*, 2008, p. 4). He also dismissed the disturbing the peace charge, ruling "The Massachusetts disturbing the peace statute cannot reach conduct which involves the exercise of a First Amendment right" (*Massachusetts v. Glik*, 2008, p. 3).

When the police returned the videographer's cell phone, some of the videos he had recorded had been erased; only one short clip remained. The flash drive had also been tampered with; the videographer incurred expenses in attempting to restore the data (*Glik v. Cunniffe et al.*, 2010a, p. 8)

A month after his arrest, the videographer filed a complaint with the Boston PD's Internal Affairs, and contacted them again after the charges against him were dismissed. The department did not investigate his complaint or take any disciplinary action against the three police (*Glik v. Cunniffe et al.*, 2010a, pp. 6-7; *Glik v. Cunniffe et al.*, 2011e, p. 4). A police memo signed by IAD Sgt. Marwan Moss in February 2008, after IAD investigators interviewed Glik, stated that the police had done nothing wrong: "Mr. Glik did not articulate a violation of law or the department's rules and regulations by an officer. Mr. Glik was advised that the proper forum for this matter was with the courts" (Cramer, 2012).

The variable of the **compensation of the videographer** is $170,000 for damages and legal fees (Ott, 2012).

The variable of the **affiliation of the videographer** is that he was a recent top-of-his-class graduate of the New England School of Law, and he had completed his recent clerkship with the Massachusetts Probate and Family Court Department, where he had also developed some working familiarity with the courts. The videographer had a future career as an attorney to lose by a wrongful felony conviction (ACLU, 2010; Frank, 2008; Volokh, 2008).

The variable of **police misconduct** is positive. It is evident from the video that police may have used excessive force (Glik, 2007). An IAD investigation later concluded that two of the police used "unreasonable judgment'' during the incident (Cramer, 2012).

The variable of the **initial response of the law enforcement agency regarding the officers involved** was that one month after his arrest, the videographer filed a complaint with the Boston PD's Internal Affairs, and followed up after his charges had been dismissed. However, the department did not investigate his complaint or take any disciplinary action against the three police (*Glik v. Cunniffe et al.*, 2011e, p. 4; *Glik v. Cunniffe et al.*, 2010a, pp. 6-7). A police memo signed by IAD Sgt. Marwan Moss in February 2008, after IAD investigators interviewed Glik, stated that the police had done nothing wrong: "Mr. Glik did not articulate a violation of law or the department's rules and regulations by an officer. Mr. Glik was advised that the proper forum for this matter was with the courts" (Cramer, 2012).

The variable of the **sequence of official actions regarding the officers involved** is complex. On February 1, 2010 the ACLU of Massachusetts filed a 42 U.S.C. § 1983 civil rights suit on Glik's behalf against the city and the three police, for violations of the videographer's First and Fourth Amendment rights, as well as state-law claims under the Massachusetts Civil Rights Act, Mass. Gen. Laws ch. 12, § 11I, and for malicious prosecution (*Glik v. Cunniffe et al.*, 2010a). The defendant police were represented by the city's attorneys (*Glik v. Cunniffe et al.*, all dates). IAD chose to re-examine Glik's 2008 complaint after the lawsuit was filed (Cramer, 2012).

On June 9, the court denied a defense motion to dismiss based on the defense that police are entitled to qualified immunity. The defendant police appealed (*Glik v. Cunniffe et al.*, 2010c), which brought a temporary stay in some of the district court proceedings for the individual police, but other proceedings against the city continued (*Glik v. Cunniffe et al.*, 2010e; PACER, 2012).

Over a year later, the U.S. Court of Appeals for the First Circuit in Boston ruled that "a citizen's right to film government officials, including law enforcement officers, in the discharge of their duties in a public space is a basic, vital and well-established liberty safeguarded by the First Amendment" (*Glik v. Cunniffe et al.*, 2011e, p. 17; Waterman, 2011). The court also concluded that "Glik was exercising clearly established First Amendment rights in filming the officers in a public space, and that his clearly-established Fourth Amendment rights were violated by his arrest without probable cause" (*Glik v. Cunniffe et al.*, 2011e, p. 2). Finally, the appeals court ruled that the police were not entitled to qualified immunity from either the videographer's First or Fourth Amendment claims (*Glik v. Cunniffe et al.*, 2011e, pp. 17, 24). With the appeal stay lifted, the case resumed as of September 21, 2011.

The department attempted to keep the progress and results of their internal investigation confidential. The court did not allow the police to succeed in these efforts. On November 7, 2011, the city filed a two-page supplemental memorandum stating that IAD was still investigating the complaint the videographer filed on November 1, 2007, four years previously (*Glik v. Cunniffe et al.*, 2011g). The police response to the court's noting that "there has not been a single document created within the last eight months" (*Glik v. Cunniffe et al.*, 2011g, p. 1), was that the investigator had recently conducted additional interviews. The memo also states that because the investigation was not complete, IAD had not yet made recommendations to the police commissioner, so the commissioner had not yet issued any findings or meted out any discipline (*Glik v. Cunniffe et al.*, 2011g, pp. 1-2). The department's memo concludes with the admission that removing the confidentiality of the investigation's documents will not impair the IAD investigation at this point (*Glik v. Cunniffe et al.*, 2011g, p. 2). The court's response to this memorandum was to deny the defendants' motion to maintain confidentiality (PACER, 2012c).

On January 5, 2012, four years and three months after the incident, Glik received a letter from Superintendent Kenneth Fong of the department's Bureau of Professional Standards, which reported that IAD had found that Cunniffe and Savalis had used "unreasonable judgment'' during the incident, but that none of the three police had used excessive force against the minor subject (Cramer, 2012). IAD also reportedly did not include Hall-Brewster in the finding because he was not initially aware of the charges the other two police filed against Glik (Cramer, 2012).

On January 9, a department spokeswoman stated that the two police now face disciplinary actions ranging from oral reprimand to suspension (Cramer, 2012). This represents a reversal of the city's position in Glik's lawsuit, one which Wunsch of the ACLU characterized as "they're hanging the individual officers out to dry" (Lee, 2012a). Glik was reportedly considering an appeal of the IAD's use of force finding to the city's Community Ombudsman Oversight Panel (CO-OP), and stated that he planned to continue his lawsuit against the city and all three police (Cramer, 2012).

In a case with many parallels to this one, the US Department of Justice (2012) filed a Statement of Interest on January 10, 2012, in *Sharp v Baltimore*, a case of a civilian whose cell phone was confiscated by police, who then deleted from it both his recordings of police actions and his personal videos (Lee, 2012b). The statement begins:

> This litigation presents constitutional questions of great moment in this digital age: whether private citizens have a First Amendment right to record police officers in the public discharge of their duties, and whether officers violate citizens' Fourth and Fourteenth Amendment rights when they seize and destroy such recordings without a warrant or due process. The United States urges this Court to answer both of those questions in the affirmative. (US DOJ, 2012, p. 1)

According to the legal director of the ACLU of Maryland, this was the first statement by the DOJ on Constitutional protection of citizens' rights to record police actions with their cell phones (Fenton, 2012).

On March 5, 2012 Magistrate Judge Dein conducted a mediation and the case was settled. The settlement order of dismissal was filed the next day, and a stipulation of dismissal of all claims was filed March 29, 2012.

The variable of the **final outcome regarding the officers involved** is that two of the three police, Cunniffe and Savalis, were departmentally disciplined (Ott, 2012; Fong, 2012). The department has not disclosed further details as of this writing.

For the variable of **final outcome of the law enforcement agency regarding policy changes, if any,** there has been no admission of guilt, but a new mandatory training video for the department, based on the videographer's actions, was introduced two months after the incident (FOX 25, 2010). A spokeswoman for the commissioner stated in January 2012 that the department has issued memos and training videos on the wiretap law, including one training bulletin that states, "There is no right of arrest for public and open recordings under this statute" (Cramer, 2012).

The variable of the **affiliation of the civilian subject** is that the civilian subject is described only as a 16-year-old, therefore a minor, and further information has not been made public (Frank, 2008; Volokh, 2008; FOX 25, 2010). For that reason, the variables of **the initial official response to handling of the civilian subject**, the **sequence of official actions regarding the civilian subject**, the **compensation (if any) of the subject**, and the **final outcome regarding the civilian subject** are all not available.

The variable of **what third parties (if any) involved themselves in the legal case** is that an ACLU attorney is part of the videographer's legal team (*Glik v. Cunniffe et al.*, 2010a, p. 10). Although the Boston Police Patrolmen's Association (BPPA) has made statements supporting the police in this case, they have not formally joined as third parties; the defendants' legal representation is entirely city attorneys. *Amicus* briefs were filed by the Center for Constitutional Rights, Berkeley Copwatch, Communities United Against Police Brutality, Justice Committee, Milwaukee Police Accountability Coalition, Nodutdol for Korean Community Development, and Portland Copwatch (*Glik v. Cunniffe et al.*, 2011b).

The variable of ***cui bono?*****, or who benefits?**, in this case are videographers and anyone else who enjoys First and Fourth Amendment protection. As the appeals court observed, "the news-gathering protections of the First Amendment cannot turn on professional credentials or status" *Glik v. Cunniffe et al.*, 2011e, p. 13).

The variable of **how many times the video was viewed online** is not comprehensive, because sites other than YouTube do not release that data (Glik, 2007). However, the advocacy YouTube video by the ACLU has been viewed over 15,000 times as of this date (ACLUMASS1, 2011).

The variable of **how much press coverage the event received** is in the low triple digits, and includes international, national, regional, network, and wire service coverage; there are at least 93 citations in LexisNexis Academic News, six law review articles, and 29 citations in ProQuest Newsstand. A Google News search for +Glik +police retrieved 117 stories, including archives.

The variable of **how much Internet discussion was linked to this event** is that there were 50 comments on the ACLU YouTube video (ACLUMASS1, 2011). A Google search for +Glik +police retrieves over 8,500 blog posts, over 3,400 forum posts, and over 200 videos.

CASE STUDY X: WINTER, MCKENNA, BAKER ET AL., 2010-03-04

During the early hours of March 3, 2010, Ben Winter and another person independently videorecorded Prince George's County police beating student John J. McKenna on a sidewalk along Knox Road in College Park, MD. This case is included for direct replication (similar results for expected reasons), per Yin (2003, pp. 5, 110). The phenomenon is a police-civilian interaction in American public space during which police misconduct may have occurred, which was videorecorded by a civilian, and the videorecording was distributed online.

Research Question and Hypotheses

For Research Question 1, what is the outcome of user-generated online video on police-civilian interactions in American public space?, the outcomes of this case include: assault charges against McKenna and Donat were dropped (Semel, 2010; Sayeed, 2010); two police were indicted for first- and second-degree assault and misconduct in office (Tucker, 2011), and one was convicted of second-degree assault; a lawsuit against the police was settled for $2 million (Valcourt, 2012, October 22); police now wear ID numbers front-and-back on helmets (WJLA, 2012, January 25).

For this case, Hypothesis 1, that user-generated online video has the potential to improve accountability in police-civilian interactions in American public space, is evidently positive for this case: false charges were dropped, police have been suspended, indicted, and convicted, and department policies have been changed. In the words of private investigator Sharon Weidenfeld,

> Beatdowns by the police are something that in my work I hear about on a daily basis. Unless we have video to prove it, nobody takes these things seriously. They say a picture is worth a thousand words. A video is worth a million. (Castaneda, 2010a)

The value of Hypothesis 2, that user-generated online video is significantly different from professional video journalism in its effects on accountability in police-civilian interactions in American public space, in this case does not appear to be evaluable without conjecture.

Hypothesis 3, that there are strong motivations for police to continue to attempt to restrict civilians' First Amendment rights to photograph police in public spaces, is in this case evidently positive, as police have been investigated, suspended, indicted, and convicted.

Documents

The case data exceed 100 files, including 15 media clips from YouTube, broadcast TV, and law office press releases. The original criminal charges did not generate many court documents. Most of the early documents were news reports, op-ed columns, blog entries, forum threads, and website comments. This has been a popular case with significant media interest. Because the incident occurred in connection with a major collegiate sporting event, news updates appeared in publications such as *Sports Illustrated*, television networks such as ESPN, and on sports websites.

Relevant Variables of Interest

The variable of **whether the police were aware of the camera** is positive. Police were in direct line of sight of the hundreds of active cell phones and camcorders visibly in use (Winter, 2010; WTOP, 2010a), and one of the police exhibits awareness of the camera when he says to a videographer, "Back up, back up, please, please. You can zoom in" (Winter, 2010, 02:40; Present, 2010a).

The variable of **when the police became aware of the camera** is evidently prior to and throughout the incident (Winter, 2010; WTOP, 2010a).

The variable of whether there was any marked change in police behavior once they were aware of the camera is negative (Winter, 2010; WTOP, 2010a).

The variable of **whether police made any attempt to prevent the recording** is apparently negative. The only evident contact between the videographer and the police is when one of the police says to Winter, "Back up, back up, please, please. You can zoom in" (Winter, 2010, 02:40; Present, 2010a). Winter was not approached by police during the crucial 30 seconds of the video; he was by then recording from an upper story of a dormitory building (Winter, 2010; WTOP, 2010a). For the same reason, the variables of **whether police made any attempt to acquire, confiscate, or destroy the video**, of **whether the police gave any unlawful instruction to the videographer regarding the video**, and of **whether police detained, cited, or arrested the videographer** are all apparently negative (Winter, 2010; WTOP, 2010a; Present, 2010a).

The variable of **when the video was available via the Internet** is March 4, 2010, the day after the incident. Winter shot the nine minute, high-definition video from his dorm room window using his new Canon EOS 7D DSLR camera, and uploaded the video to his YouTube channel, frozenphoenixprod, where it is still available at full quality, significantly better than the broadcast versions (Winter, 2010; Present, 2010a). Winter then responded to an almost immediate query from private detective Sharon Weidenfeld, who had been retained by McKenna's attorney, Chris Griffiths. The attorney and private detective released a 30-second, stabilized and cropped clip from the video as part of a press release after the charges against McKenna and Donat were dropped (Present, 2010a; Sayeed, 2010). The shorter video was broadcast (WTOP, 2010b; Associated Press, 2010a) and uploaded in a variety of forms to YouTube and other websites on and after April 12 (o1OpTiMuS1o, 2010). Winter's video has been supplemented by cell phone video from at least one videographer at street level (WTOP, 2010a). Throughout Winter's footage, it is evident that a significant fraction of the crowd are taking pictures and video with their cell phones (Winter, 2010).

The variable of **when the video was available via broadcast news media** is that McKenna's attorney released the video as part of a press release on April 12 (Semel, 2010); the video was broadcast the same day (o1OpTiMuS1o, 2010).

The variable of when images from the video were available via print news media is on April 13 (Calvert & Jones, 2010).

The variable of **whether more than one camera captured the event** is positive. Winter's video has been supplemented by cell phone video from at least one videographer at street level (WTOP, 2010a). Throughout Winter's footage, it is evident that a significant fraction of the crowd are taking pictures and video with their cell phones (Winter, 2010).

The variable of **whether official CCTV, dashboard, or other video captured the event** is disputed. The University of Maryland has 350 CCTV cameras on campus; camera 158 covers Knox Road from South Campus Commons buildings 3 (Winter's location) and 4 up to Route 1, the area where the incident took place (Present, 2010b). Footage from camera 158 for the time of the incident has been reported as missing or damaged, but there are discrepancies in the official statements. Other CCTV footage from the same day, at first reported missing, has later come to light.

The variable of **whether police initially admitted to the existence or possession of video of the event** is negative. Police initially denied the existence of CCTV or police video of the event (Present, 2010b).

The variable of **when did police admit to possession of video of the event** is complex. During the discovery phase of pretrial investigation in March and early April 2010, campus security recordings were located and approximately 60 hours of video was turned over to McKenna's attorney, Chris Griffiths, in compliance with a subpoena (Present, 2010b; WTOP, 2010c). University police technician(s) were responsible for removing video from multiple cameras; recordings from 15 other cameras were delivered to McKenna's attorney, but the recording from camera 158 was omitted. According to Griffiths, "it was the one camera — the most important camera — that was omitted. They say it's a mistake, but it's a coincidence which raises enough questions that one would hope that it's thoroughly investigated" (Present, 2010b).

Campus police Lt. Joanne Ardovini was in charge of monitoring the campus CCTV system on March 4 (Present, 2010b; WTOP, 2010c). She is married to one of the Maryland National Capital Park Police mounted unit officers, John Ardovini, who was on duty during the incident (WTOP, 2010c; Present, 2010b). He was named in the charging documents against McKenna and Donat, and is one of the police who was suspended (WTOP, 2010c; Present, 2010b; WJLA, 2010).

A copy of the missing recording was turned in April 20 by Lt. Jim Goldsmith, commander of the University Police investigative unit. He reportedly had been conducting his own investigation into the riot, and had made a copy of the video from camera 158. When Goldsmith learned the footage was missing, he turned his copy over to officials. According to campus officials, because the campus video system has a fixed capacity, and any recording that is not duplicated is eventually overwritten, the original footage would have been permanently lost without Goldsmith's copy (Present, 2010b; WTOP, 2010c). However, when Goldsmith's copy was examined, there were still three minutes of video missing (WTOP, 2010c; Present, 2010b).

On April 21, University spokesman Milree Williams and University Police spokesman Paul Dillon held a press conference in which they said state police had been asked to investigate the matter and to review university police procedures (WTOP, 2010c; Present, 2010b). Both officials stated they don't believe any employee misconduct took place. Williams also said that Ardovini had removed herself from the investigation over the potential conflict of interest (Present, 2010b).

The limited information released after the state police investigation states that Lt. Ardovini recused herself from the investigation before the recording went missing, that the technician Lt. Ardovini would have supervised made an oversight in omitting the recording, and that the missing several minutes would not have showed the incident because the technician monitoring that camera had it pointed elsewhere at the time. Dillon stated, "It was an honest mistake" (Habtemariam, 2010).

The variable of **whether police released official video of the event** is, police have not released any official video of the incident to the public, but they have released video to McKenna's attorneys in compliance with a subpoena (Present, 2010b).

The variable of whether official video of the event was available via the Internet is negative.

The variable of **who was credited as the source for each medium of release of the video** is, video sources have variously been credited as YouTube, WJLA, and Associated Press; Winter has not evidently been credited by name in any medium of release other than the campus newspaper, *The Diamondback* (Associated Press, 2010a; WTOP, 2010b; o1OpTiMuS1o, 2010; Roberts & Wood, 2010; Present, 2010a).

The variable of **whether there was any effort to restrict, remove or prosecute the release of the video** is negative, based on the document search and acquisition procedures detailed previously. The civilian videographer has experienced no attempts to restrict or prosecute the release of his video, and there has been no official response to the videographer at all (Winter, 2010). For the same reasons, the variables of the **initial official response to handling of the videographer**, of the **sequence of official actions regarding the videographer**, and of the **final outcome regarding the videographer** are all not applicable.

The variable of the **compensation of the videographer** is unknown, based on the document search and acquisition procedures detailed previously.

The variable of the **affiliation of the videographer** is that at the time of the incident he was a senior electrical engineering major at the University of Maryland; he has no known affiliation with professional journalism, police, or activist groups, although he is an amateur filmmaker who has had some local success and recognition (Present, 2010a). He also stated to the campus newspaper,

> P.G. County Police clearly have a history of brutality and misconduct, and I think anyone who thinks that this video is going to solve all those problems is kind of missing the forest for the trees. This is an event that will likely be repeated if something serious and permanent is not instituted. (Present, 2010a)

The variable of **whether police misconduct was recorded** is positive. Police misconduct of excessive force was recorded, and police were indicted (Winter, 2010; Tucker, 2011) and convicted (Valcourt, 2012, October 22).

The variable of the **initial response of the law enforcement agency regarding the officers involved** was suspension of several police pending internal investigation. "Officer Sean McAleavey was suspended Monday [April 12], just hours after Winter's video made national news, and an unidentified sergeant was suspended Tuesday evening [April 13]" (Present, 2010a). The sergeant was later identified as Anthony Cline (Present, 2010c).

The police chief stated that the officers did not file a use of force report, that he expected more suspensions as the other officers in the incident were identified, and that he was "very disappointed with what I saw" (Sayeed, 2010) when he viewed the video on April 12. The chief also stated that the case was an "isolated incident" (Noble, 2010). For context, the department was just one year out from under federal oversight. For six years prior to that, the Department of Justice had monitored the department and mandated changes, based on previous patterns of police brutality (Noble, 2010).

The variable of the **sequence of official actions regarding the officers involved** is complex. By April 15, police spokesman Maj. Andrew Ellis said officials had identified all the officers involved in McKenna's beating, and that "Criminal charges are a real possibility" (Present, 2010a). Ellis also explained that three investigations were in process: an internal investigation as to possible violations of policies or procedures, a criminal investigation that could lead to police being charged, and a civil rights investigation (Associated Press, 2010b). Evidently, none of the police who had contact with McKenna filed a use-of-force report, an infraction of department regulations (Castaneda, 2010b).

By the end of April, four police had been suspended or "placed on administrative leave" over the false charging documents: Officers Reginald H. Baker, James Harrison, Jr., Sean McAleavey, and Sgt. Anthony J. Cline (Present, 2010c; Tucker, 2011).

In late June, police internal affairs detectives working for the state's attorney's office reviewed the email and cell phone messages of police commanders during the night of the incident and in mid-April, when the video made national headlines (Castaneda, 2010b). Major Kevin Putnam was in command of approximately 100 police in riot gear, including Baker and Harrison (Castaneda, 2010b).

As of July 19, Baker, Cline, Harrison, and McAleavey were still reportedly suspended with pay (Castaneda, 2010b).

By early October, the police internal affairs investigation was mostly complete, but was "put on hold" at the request of the federal authorities, according to Ellis (Zapotosky & Castaneda, 2010; Wagner, 2010).

Nine months after the incident, the FBI and the Justice Department's Civil Rights Division took over the investigation, beginning with unannounced knock-and-talk interviews of 40 officers in their homes the evening of December 2 and continuing into December 3 (Zapotosky & Castaneda, 2010; Wagner, 2010). Baker, Cline, and Harrison remained suspended or on desk duty at the time; McAleavey had been returned to duty. FBI spokesman Richard Wolf said that federal authorities had been monitoring the local investigations, and were now investigating the incident as a civil rights case (Zapotosky & Castaneda, 2010).

On September 20, 2011, State's Attorney Angela Alsobrooks announced indictments charging Baker and Harrison with first- and second-degree assault and misconduct in office (Broom, 2011). McAleavey was not named in the indictment. Officials were reportedly still trying to determine if he knowingly signed a false police report (Giles, 2011).

Baker and Harrison had been on paid administrative leave since the incident (Giles, 2011). They turned themselves in September 21, and were processed and released on $75,000 bond (Gordon & Stabley, 2011). The Fraternal Order of Police Lodge 89 issued a statement crediting the two police with "long and exemplary" careers, and stated, "We believe that it would be irresponsible and unfair to rush to judgment" (Gordon & Stabley, 2011). The indictment was handed down after a 16-month investigation.

The variable of the **final outcome regarding the officers involved** is complex. McAleavey has been returned to duty, but remains under investigation (Giles, 2011). There is an ongoing federal civil rights investigation (Zapotosky & Castaneda, 2010). Sgt. Jones, who was named as being assaulted by Donat, has since retired. Baker and Harrison were indicted and out on bond (Gordon & Stabley, 2011). At trial, Baker was cleared of all charges (Associated Press, 2012, October 19), but Harrison was convicted of second-degree assault and sentenced to 30 days home detention and 18 months probation, and retired "with prejudice" (Zapotosky, 2012, December 14).

For the variable of **final outcome of the law enforcement agency regarding policy changes, if any**: No final policy changes are yet public. However, it is important to consider the position of the police department in the political structure of Prince George's County. The chief of police is an appointed position; Hylton, chief at the time of the incident, had been appointed in 2008 by then-County Executive Jack Johnson, who later pled guilty to corruption charges. Hylton was removed in December, 2010, by one of the first acts of the new executive, and Magaw was named interim chief. Magaw was confirmed as chief in early July, 2011. In contrast, the state's attorney is an elected position. Alsobrooks was sworn in January 3, 2011. The investigations for this incident have therefore seen significant turnover at the highest levels, and the criminal prosecution has significant political implications for the department and the county government.

The variable of the **affiliation of the civilian subject** is that at the time of the incident he was a 21-year-old junior at the University of Maryland (Present, 2010a). Notably, McKenna comes from a family of lawyers, and is the grandson of a retired circuit court judge (Noble, 2010). The second civilian subject assaulted by police was 19-year-old sophomore Benjamin Donat.

The variable of **the initial official response to handling of the civilian subjects** is complex. The videos show a crowd in the street. McKenna jogs down the sidewalk; in the cell phone video he is audibly singing a cheer (WTOP, 2010a). McKenna stops in front of two mounted police, who corner him against a wall. Three police in riot gear run at him and slam him into the wall; he then falls to the sidewalk. One of the police turns and moves further down the sidewalk, and the remaining pair continue to beat McKenna with batons while he is prone on the sidewalk (Winter, 2010; WTOP, 2010a, 2010b; Associated Press, 2010a).

According to charging documents, police then transported McKenna and another student, Benjamin Donat, to Upper Marlboro for processing (WJLA, 2010).

After their interactions with police, McKenna and Donat both had concussions, cuts and other injuries (WTOP, 2010c). According to McKenna's attorney, the charging documents omit several important facts about these injuries: McKenna was treated by EMTs at the scene of the incident, which requires separate police paperwork that was not submitted; in transit, police removed the bandage from McKenna's head, and told him that if he said anything about his injuries, he would be held over the weekend rather than released; and the jail personnel noted that McKenna was bleeding, and insisted that the arresting officers take him to the hospital (ABC News, 2010). The injuries sustained by McKenna (including the eight staples closing the cut in his scalp) were later photographed by the investigator working for his attorney, and those images were made available on the Internet and edited into some of the television broadcasts and online videos (WTOP, 2010c; Noble, 2010).

The variable of the **sequence of official actions regarding the civilian subjects** was complex. Police charged McKenna with second-degree assault of Ardovini (whose name was misspelled on the charging documents, causing much confusion in news reports), and charged Donat with second-degree assault of Jones; both officers were members of the Maryland National Capital Park Police (WJLA, 2010). Both students were also charged with disturbing the peace (WJLA, 2010). The charging documents filed by the police were made public (WJLA, 2010). The Statement of Probable Cause on McKenna's reads (*sic*):

> Arrested 1 (Mckenna, John James) and Arrested 2 ([Donat, Benjamin]) were running in the middle of Baltimore avenue screaming. Due to thir disorderly behavior a crowd on the sidewalk began to form and become unruly. As Officer Ardozini #246 and Officer Jones #177 from the maryland National Capital Park Police mounted unit attempted to regain order, Arrested 1 and arrested 2 struck those officers and their horses causing minor injuries. Arrested 1 and Arrested 2 were both kicked by the horses and sustained minor injuries. (WJLA, 2010)

The arresting officer is listed as McAleavey, P.G. County Police ID No. 3052, and the signature is dated 3/4/10.

None of what are stated as facts in the charging document match what can be observed in the two videos (Winter, 2010; WTOP, 2010a; WJLA, 2010). According to private investigator Weidenfeld, Donat was injured a block away from the site of McKenna's beating; the two students did not know each other, and were not together at the time of the incident (Semel, 2010; Noble, 2010).

A spokesman for Prince George's County State's Attorney Glenn Ivey said that prosecutors dropped the charges against McKenna due to a lack of evidence before they saw the video (Sayeed, 2010). According to the students' attorney, on April 12 the charges against McKenna were dropped without comment, and on April 9 a prosecutor stated that charges against Donat were dropped because officers could not identify him (Semel, 2010).

The variable of the **compensation (if any) of the subject** is that the civilian subject's civil rights lawsuit was settled for $2 million (Valcourt, 2012, October 22).

The variable of the **final outcome regarding the civilian subjects** is that the civilian subject's civil rights lawsuit was settled (Valcourt, 2012, October 22).

The variable of **what third parties (if any) involved themselves in the legal case** is none, based on the document search and acquisition procedures detailed previously.

The variable of ***cui bono?*, or who benefits?**, in this case is the civilian subject, the nine other students whose civil suit was settled for a total of $3.6 million, and their attorneys (Valcourt, 2012, October 22).

The variable of **how many times the video was viewed online** is, in aggregate, over 900,000 views on YouTube alone (Winter, 2010; Associated Press, 2010a; WTOP, 2010b; WTOP, 2010a; o1OpTiMuS1o, 2010). This does not include the views from AP or broadcast station websites.

The variable of **how much press coverage the event received** included 29 stories from LexisNexis Academic News and 39 from ProQuest Newsstand. A Google News search for +McKenna +police +Maryland +Duke retrieved 119 stories, including archives. Coverage was international, via UPI and AP wire services, and all broadcast networks. There was a cluster of new stories when two of the police were indicted, during their trial, and at sentencing.

The variable of **how much Internet discussion was linked to this event** is that there were over 4,000 comments among the leading YouTube videos of the incident. A Google search for +McKenna +police +Maryland +Duke retrieved over 2,300 videos, nearly 18,000 blog posts, and almost 3,000 forum posts.

CASE STUDY XI: WILLIAMS ET AL., CHAPMAN ET AL., 2009-08-20

On the night of August 20, 2009, on a public sidewalk in Brooklyn, Taneisha Chapman and Markeena Williams refused to show identification to NYPD Officer Eugenia Williams and Sergeant Marshal Winston, who arrested the two civilians without probable cause. This case is included for contrast (differing results for expected reasons), per Yin (2003, pp. 5, 110); there is no mention of a camera, but otherwise the phenomenon is a police-civilian interaction in American public space during which police misconduct may have occurred.

Research Question and Hypotheses

For Research Question 1, what is the outcome of user-generated online video on police-civilian interactions in American public space?, the outcomes of this case include: a $20,000 settlement to each of the two civilians (these amounts included attorney's fees), and no admission of guilt by the individual police, the department, or the city (Chapman et al., 2011 June 6, p. 2).

For this case, Hypothesis 1, that user-generated online video has the potential to improve accountability in police-civilian interactions in American public space, is not evaluable without cross-case analysis.

The value of Hypothesis 2, that user-generated online video is significantly different from professional video journalism in its effects on accountability in police-civilian interactions in American public space, in this case does not appear to be evaluable without conjecture.

Hypothesis 3, that there are strong motivations for police to continue to attempt to restrict civilians' First Amendment rights to photograph police in public spaces, in this case does not appear to be evaluable without conjecture.

Documents

Documents for this case include a single report in one newspaper (Marzulli, 2010), plus mentions within stories on NYPD stop-and-frisk practices in two other newspapers (Hentoff, 2010; Rivera, et al., 2010) and two blogs (Saxena, 2010; Cummings, 2010). There are 31 data files, including one ACLU video overview of police stops (ACLU &

White, 2011), print media stories, reader comments (particularly EducatedBlackGirl, 2010), ACLU palm cards and flyers (NYCLU, 2004; ACLU, n.d.b), and the seventeen documents filed in the civil suit, six of which contain relevant data (Chapman et al., all dates).

Relevant Variables of Interest

By the nature of this case, there are no data for the **24 variables of interest regarding civilian cameras**, **videographer**, or **video**. Furthermore, examination of Google Street View images of the location of the incident did not reveal any indications of CCTV cameras on the exteriors of any of the nearby buildings, either the Marcy Houses towers or the buildings on the other side of Nostrand Avenue. Only fifteen of New York City Housing Authority's 334 developments have CCTV cameras that are monitored by police 24 hours a day, seven days a week (Reynolds, 2010).

The variable of **police misconduct** is positive; the officers made false arrests, although the terms of the settlement enable the police to not admit any wrongdoing (Chapman et al., 2011 January 10, pp. 1, 3-10; Chapman et al., 2011 June 6, p. 2).

Based on the document search and acquisition procedures, there are no public records for this case for either the variable of **initial response of the law enforcement agency regarding the officers involved**, or of **sequence of official actions regarding the officers involved**, or of **final outcome regarding the officers involved**; there has been no admission of guilt, and no evidence of internal investigation. For the variable of **final outcome of the law enforcement agency regarding policy changes, if any**: there has been no evidence of any departmental policy changes, and the NYPD's stop-and-frisk tactics, which led to this incident, remain controversial and are the subject of ongoing protests and at least one class-action lawsuit. These tactics have also been the subject of legal and government policy analysis in recent years (La Vigne et al., 2012; Zeidman, 2013; Harvard Law Review, 2013a; Lerman & Weaver, 2014).

The variables of the **affiliations of the civilian subjects** appear to be entirely civilian; neither one is a journalist, nor are they affiliated with any media organization or with the police, or with any activist groups known for provoking the police. However, Markeena Williams identified herself in an online forum as having a criminal justice degree, and of pursuing a law degree (EducatedBlackGirl, 2010).

The variable of **initial official response to handling of the civilian subjects** is “unspecified criminal charges” (Marzulli, 2010) due to the inability of the district attorney’s office to locate any record of the civilians’ arrests. According to the civil suit’s amended complaint, police

> told the Kings County District Attorney’s Office that plaintiffs had committed various crimes; based upon the false statements of defendants, the Kings County District Attorney’s Office prosecuted plaintiffs; both the prosecutions concluded in adjournments in contemplation of dismissal on the date of arraignment, and were eventually dismissed. (Chapman et al., 2011 January 10, p. 6).

The variable of the **sequence of official actions regarding the civilian subjects** was to: arrest and handcuff the civilians; transport them to the PSA-3 Housing Precinct; transfer them to Brooklyn Central Booking; hold them for arraignment; arraign them on unspecified criminal charges; adjourn the prosecution in contemplation of dismissal (ACD); release the civilians on their own recognizance; automatically dismiss the charges after the prosecution failed to present evidence (Chapman et al., 2011 January 10; Marzulli, 2010; Saxena, 2010; Hentoff, 2010; Cummings, 2010); file a series of categorical denials and delaying legal motions in the civilians’ civil rights suit, until the end of the written discovery period on May 19, 2011 (PACER, 2011); and then to settle the suit before trial (Chapman et al., 2011 June 6, p. 2).

The variable of the **final outcome regarding the civilian subjects** was the payment of $20,000 each, that amount to include the plaintiff’s attorneys’ fees (Chapman et al., 2011 June 6, p. 2).

The variable of **third parties to the legal case** is none; however, the ACLU prominently featured the case on its website and blog (Cummings, 2010).

The variable of ***cui bono?***, or who benefits, in this case is the two civilians and their attorneys (Chapman et al., 2011 June 6, p. 2), at least monetarily.

The variable of **press coverage linked to the event** included a single report in the *New York Daily News* (Marzulli, 2010) following the filing of the civilians' civil rights suit, plus mentions within NYPD stop-and-frisk practices stories in the *Village Voice* (Hentoff, 2010) and the *Gothamist* blog (Saxena, 2010). There was no follow-up press coverage when the case was settled, based on the document search and acquisition procedures. The variable of **Internet discussion linked to this event** is in the low single digits, almost all of which is in the form of comments posted to the websites where the aforementioned press coverage was published.

CASE STUDY XII: BUSHWICK 32, NYPD, 2007-05-01

On the afternoon of May 1, 2007, as they walked on a public sidewalk in Brooklyn on their way to a wake, 32 young civilian men and women ranging in age from 13 to 22 were arrested by the NYPD without probable cause. This case is included for contrast (differing results for expected reasons), per Yin (2003, pp. 5, 110); the event was not captured in its entirety by a civilian camera, but otherwise the phenomenon is a police-civilian interaction in American public space during which police misconduct may have occurred.

Research Question and Hypotheses

For Research Question 1, what is the outcome of user-generated online video on police-civilian interactions in American public space?, the outcomes of this case include: $447,500 in total settlements among 30 of the civilians (this amount included attorney's fees), and no admission of guilt by the individual police, the department, or the city (*Prosper et al v. City of New York et al.*, 2009 May 12; *Jackson et al v. City of New York et al.*, 2010 March 2; *Green, Jr. et al v. The City of New York et al.*, 2009 July 6; *Chavarria et al v. City of New York et al.*, 2009 June 24).

For this case, Hypothesis 1, that user-generated online video has the potential to improve accountability in police-civilian interactions in American public space, is not evaluable without cross-case analysis.

The value of Hypothesis 2, that user-generated online video is significantly different from professional video journalism in its effects on accountability in police-civilian interactions in American public space, in this case does not appear to be evaluable without conjecture.

Hypothesis 3, that there are strong motivations for police to continue to attempt to restrict civilians' First Amendment rights to photograph police in public spaces, in this case does not appear to be evaluable without conjecture.

Documents

Documents for this case include: 85 court documents distributed among the four civil rights lawsuits, although only a few documents for each court case contain data useful for determining the variables of interest; the criminal complaints sworn out by the police to the district attorney, and appended as exhibits to the complaints for each of the four lawsuits; over 30 print media reports, drawn from *The New York Times*, *New York Daily News*, *Village Voice*, *Gothamist*, *Amsterdam News*, and a scattering of smaller news outlets; three radio and two television broadcasts; and entries from a handful of blogs and online forums. Notably, only a small fraction of the media reports on this case are from mainstream news organizations.

Relevant Variables of Interest

The variable of **whether the police were aware of the camera** is apparently negative, based on the document search and acquisition procedures. There do not appear to be any publicly available images of the beginning of the incident. The only available images of the event were taken after the civilians were handcuffed, by an eyewitness with a cell phone camera taking photos (Herbert, 2007c), and by a helicopter recording video from overhead (mpizzie, 2009). The variable of **when the police became aware of the camera** is after the civilians were handcuffed, and a resident of the adjacent building began to take pictures with his cell phone camera and to ask the police why the group was being arrested (Herbert, 2007c). The variable of **whether there was any marked change in police behavior once they were aware of the camera** is not applicable, as there is no before-and-after video for purposes of comparison.

The variables of whether police made any attempt to prevent the recording, whether police made any attempt to acquire, confiscate, or destroy the video, whether the police gave any unlawful instruction to the videographer regarding the video, and whether police detained, cited, or arrested the videographer are all apparently negative, as the eyewitness with the cell phone camera did not report any such police actions (Herbert, 2007c) and the helicopter video was broadcast that day (mpizzie, 2009).

The variable of **when the video was available via the Internet** is April 21, 2009, as part of an edited video of the Bushwick 32 story released when the first settlements were announced (mpizzie, 2009). The variable of **how many times the video was viewed online** is 413 (mpizzie, 2009). The variable of **when the video was available via broadcast news media** is May 21, 2007 for the original helicopter video, in a segment entitled, "33 Arrested in Funeral Fight" (mpizzie, 2009). The variable of **when images from the video were available via print news media** is apparently never, based on the document search and acquisition procedures; none of the available print media have included any video images.

The variable of **whether more than one camera captured the event** is positive. Public documents reveal at least two: the cell phone camera used by an eyewitness and the video camera in the helicopter (Herbert, 2007c; mpizzie, 2009).

The variable of **whether official CCTV, dashboard, or other video captured the event** is unknown, based on the document search and acquisition procedures. There were multiple police vehicles, and the captain in charge had been monitoring the park as the group gathered, but police never produced any video of the event (Herbert, 2008). For the same reason, the variables of **whether police initially admitted to the existence or possession of video of the event**, **when did police admit to possession of video of the event**, **whether police released official video of the event**, and **whether official video of the event was available via the Internet** are all not applicable.

The variable of **who was credited as the source for each medium of release of the video** is that the YouTube video cropped out the station ID, so there was no visible credit (mpizzie, 2009). The variable of **whether there was any effort to restrict, remove or prosecute the release of the video** is evidently negative, based on the document search and acquisition procedures and the fact that the video was broadcast. The variable of the **initial official response to handling of the videographer** is not applicable, as the police at the scene did not interact with the helicopter camera operator. For the same reasons, the variables of the **final outcome regarding the videographer** and the **sequence of official actions regarding the videographer** are not applicable. The variable of the **compensation of the videographer** is a matter of his or her employment, as is the variable of the **affiliation of the videographer** as a professional journalist.

The variable of **police misconduct** is positive; the officers made false arrests, although the terms of the settlement enable the police to not admit any wrongdoing (*Prosper et al v. City of New York et al.*, 2009 May 12; *Jackson et al v. City of New York et al.*, 2010 March 2; *Green, Jr. et al v. The City of New York et al.*, 2009 July 6; *Chavarria et al v. City of New York et al.*, 2009 June 24).

The variable of **initial response of the law enforcement agency regarding the officers involved**, was for NYPD chief spokesman Paul J. Browne to publicly defend the police actions: "The captain, a high-ranking official in the department, an experienced police captain, made a good-faith judgment and ordered the arrests" (Lee, 2007a).

The variable of **sequence of official actions regarding the officers involved** is that the Brooklyn district attorney's office opened its own investigation. Chief spokesman Jerry Schmetterer told reporters, "We've heard from the community, and there have been a number of questions to the office as to what happened" (Lee, 2007a). A few weeks later, Brooklyn DA Charles Hynes stated on WNYC that his office had conducted an "independent inquiry" and that

> We had many, many interviews with local store owners and people who live in the neighborhood who are, frankly, scared to death of these kids. And they were not just walking on one car; they were trampling on all sorts of cars. It was almost as if they were inviting their arrest. (Herbert, 2008)

To the *Daily News*, Hynes added, "It began to erupt into a full-scale disturbance, with kids also blocking traffic and blocking pedestrian walks" (... and uncovering lies, 2007). Based on the document search and acquisition procedures, there have been no statements on the incident from the NYPD Internal Affairs Bureau, nor have there been any official actions regarding the unsubstantiated and contradictory statements sworn to by police in the charging documents (*Jackson et al v. City of New York et al.*, 2009 June 22, pp. 24-36).

Based on the document search and acquisition procedures, there are no public documents for this case for the variable of **final outcome regarding the officers involved**; there has been no admission of guilt, and no evidence of internal investigation (*Prosper et al v. City of New York et al.*, 2009 May 12; *Jackson et al v. City of New York et al.*, 2010 March 2; *Green, Jr. et al v. The City of New York et al.*, 2009 July 6; *Chavarria et al v. City of New York et al.*, 2009 June 24). For the variable of **final outcome of the law enforcement agency regarding policy changes, if any**: there has been no evidence of any departmental policy changes, and the NYPD's stop-and-frisk tactics, which led to this incident, remain controversial and are the subject of ongoing protests and at least one class-action lawsuit. These tactics have also been the subject of legal and government policy analysis in recent years (La Vigne et al., 2012; Zeidman, 2013; Harvard Law Review, 2013a; Lerman & Weaver, 2014).

The variable of the **affiliations of the civilian subjects** appear to be entirely civilian; none of them were journalists, nor were they affiliated with any media organization or with the police, or with any activist groups known for provoking the police. Police were unable to substantiate any of their statements that the civilians were wearing gang colors such as red bandanas (Herbert, 2007c). Most of the civilians were students at Bushwick Community High School, where teacher Brian Favors said, "Some of these kids are our leaders. These kids are not gangsters, but you can't live in the hood and not have a cousin or brother or relative that is associated" (Lee, 2007a).

The variable of **initial official response to handling of the civilian subjects** was to detain, search, handcuff, and arrest them (Burke & White, 2007; Herbert, 2007a; Hogarty, 2009; Lee, 2007a; mpizzie, 2009; *Prosper et al v. City of New York et al.*, 2008, August 14; *Jackson et al v. City of New York et al.*, 2009 June 22; *Green, Jr. et al v. The City of New York et al.*, 2008 June 4; *Chavarria et al v. City of New York et al.*, 2008 July 23).

The variable of the **sequence of official actions regarding the civilian subjects** is complex, and documents present conflicting data. Police reported that the 83rd precinct acted on a call from members of Brooklyn's Community Board 4 (CB4) that, fearing that enemies of the deceased might attack his friends, asked the police to intervene for the group's safety when it gathered at a neighborhood park. CB4's district manager was unable to confirm or deny what any of the 15 board members might have done individually, but the board did not apparently make an official request of the police (Hutton, 2007; Lee, 2007a). According to Browne,

> Captain Scott Henderson ... who happens to be black, personally patrolled the vicinity of Putnam Park on Monday where he observed groups of six to eight individuals each, with red bandanas, and making gang signs with their hands, converge on Putnam Park at a wall that had been tagged with gang symbols. The group grew in size to 32, some wearing gang bandanas, and all wearing T-shirts memorializing McFarland. They proceeded on foot toward the 'L' train to attend a wake for McFarland in the 60th Precinct. En route, they took the entire sidewalk and part of the street; and some walked on the tops of parked cars as the group proceeded. (Hutton, 2007)

The charging documents drawn up by the district attorney's office, which bear warnings that "False statements made in this document are punishable as a Class A misdemeanor pursuant to Section 210.45 of the Penal Law," include statements by arresting officers that "based upon deponents training and experience the above mentioned defendants and said apprehended others are gang members…and were engaged in gang activity" (*Jackson et al v. City of New York et al.*, 2009 June 22, pp. 24-25). In addition, the officers stated that they observed members of the group say "Fuck the police," obstruct pedestrian traffic, obstruct vehicular traffic, and alarm and annoy pedestrians (*Jackson et al v. City of New York et al.*, 2009 June 22, pp. 24-25). Police were later unable to produce the gang bandanas as evidence, and multiple civilian eyewitnesses from the park and streets denied that any of the group had walked on cars, were misbehaving in any way, or had blocked the street or sidewalk (Herbert, 2007c).

In an orchestrated maneuver led by Henderson, police from the 83rd Precinct arrived in four police wagons, four radio cars, and two unmarked cars, came at the group with guns drawn, and frisked, handcuffed and arrested everyone in the group. According to Henderson's report, six females were issued summonses for disorderly conduct and six other people were cited as juveniles; they were reportedly released from the 83rd Precinct within two hours of their arrests. The males were transferred to Brooklyn Central Booking and held, some for up to 36 hours; their T-shirts were confiscated as evidence. Police questioned those held for information about gang activities. They were arraigned the next morning on disorderly conduct and unlawful assembly charges. Two were held on prior charges, one for marijuana possession (Hutton, 2007; Burke & White, 2007; Herbert, 2007a; Hogarty, 2009; Lee, 2007a; mpizzie, 2009; *Prosper et al v. City of New York et al.*, 2008, August 14; *Jackson et al v. City of New York et al.*, 2009 June 22; *Green, Jr. et al v. The City of New York et al.*, 2008 June 4; *Chavarria et al v. City of New York et al.*, 2008 July 23).

Hynes offered community service assignments in return for guilty pleas (Ganging up on cops, 2007; Herbert, 2008; Lee, 2007b). None of the arrestees accepted the DA's offer.

Prosecutors did not bring any cases to trial; nine months later, ten cases were still pending (Herbert, 2008). Eventually, all the cases were dismissed for lack of evidence. For example, an assistant DA moved to have the charges against one defendant dismissed after his lawyer filed a motion demanding that the DA's office produce documentary evidence of the defendants misbehaving. No evidence was ever produced of the arrestees blocking traffic or climbing on cars (Herbert, 2008; Chung, 2008; Newman, 2008). The DA's office was evidently successful in getting one arrestee to accept a plea bargain over prior charges (Kellner, 2009; Dwoskin, 2009b).

The official action to the civilians' civil rights lawsuits was to file a series of categorical denials and delaying legal motions, and then to settle the suits before trial (*Prosper et al v. City of New York et al.*, all dates; *Jackson et al v. City of New York et al.*, all dates; *Green, Jr. et al v. The City of New York et al.*, all dates; *Chavarria et al v. City of New York et al.*, all dates; PACER, all dates).

The variables of the **final outcome regarding the civilian subjects** and the **compensation (if any) of the civilian subjects** was a settlement for $447,500 in amounts of $23,000 for one plaintiff, $20,000 each for 15 plaintiffs, $9,000 for 13 plaintiffs, and $7,500 for one plaintiff, those amounts to include the plaintiff's attorneys' fees (*Prosper et al v. City of New York et al.*, 2009 May 12; *Jackson et al v. City of New York et al.*, 2010 March 2; *Green, Jr. et al v. The City of New York et al.*, 2009 July 6; *Chavarria et al v. City of New York et al.*, 2009 June 24). One plaintiff dropped out of one of the suits (PACER, 2011c), and another had accepted a plea bargain over prior charges and was not a party to the suits (Kellner, 2009; Dwoskin, 2009b).

The variable of **what third parties (if any) involved themselves in the legal case** is none. Public statements of support for the civilians were made by representatives of the groups One Hundreds Blacks in Law Enforcement Who Care and the New York Civil Liberties Union, but neither filed an *amicus* brief or otherwise participated in any of the four civil rights lawsuits (Mfuni, 2007c).

The variable of **cui bono?**, or who benefits, in this case is the 30 civilians and their attorneys (*Prosper et al v. City of New York et al.*, 2009 May 12; *Jackson et al v. City of New York et al.*, 2010 March 2; *Green, Jr. et al v. The City of New York et al.*, 2009 July 6; *Chavarria et al v. City of New York et al.*, 2009 June 24), at least monetarily.

The variable of **how much press coverage the event received** included: Two television and three radio broadcasts, and twelve print or online media reports and editorials in the month following the incident; another 18 reports over the next year, but very few in the mainstream press; and seven following the announcement of the first civil case settlements. Based on the document search and acquisition procedures, no news media have reported the total settlement. The variable of **how much Internet discussion was linked to this event** is several hundred individual messages, many of which are in the form of comments posted to the websites where the aforementioned press coverage was published. In addition, at least 355 blog entries discussed the event. A relative handful of these documents contain unique data relevant to one or more of the variables of interest. One of the factors in locating documents in this case was that the appellation chosen by the group, Bushwick 32, was not used by the mainstream media or by the courts, but was a key search term in locating relevant Internet discussion and non-mainstream media coverage. For example, *The New York Times* did not use the phrase, but the *New York Amsterdam News* did. A Google search for "Bushwick 32" returned 323 results without duplicates, and 1,970 results with duplicates.

CASE STUDY XIII: NINJA636B, RODRIGUEZ, SOUSA, 2006-09-03

On September 3, 2006, from a public sidewalk in the downtown Himmarshee bar and nightclub district of Fort Lauderdale, a civilian tourist videorecorded FLPD Sgt. Frank Sousa and Officer Zachary Baro arresting civilian Carlos Rodriguez. This case is included specifically for the negative value for the variable of whether police misconduct was or may have been recorded. It is the only such case of a civilian video reported in the news databases for the research period; all other reported videos that cleared police of misconduct were from professional news, commercial CCTV, or police cameras. The search structure used in the news databases was: ti((police OR cop? OR deput* OR sheriff?) NEAR/10 (clear* OR exonerat*) NEAR/10 (video*)). This case is included for direct replication (similar results for expected reasons), per Yin (2003, pp. 5, 110). The phenomenon is a police-civilian interaction in American public space which was videorecorded by a civilian, and the videorecording was distributed online.

Research Question and Hypotheses

For Research Question 1, what is the outcome of user-generated online video on police-civilian interactions in American public space?, the outcomes of this case include: the police department initiated an investigation of Sousa's actions based on the YouTube video, absent a complaint from Rodriguez; Sousa was cleared of any misconduct; and Rodriguez was prosecuted (ninja636b, 2007; Wallman, 2007a, b; Citizens Police Review Board, 2007a, b, c; *State of Florida v. Rodriguez*, 2007).

For this case, Hypothesis 1, that user-generated online video has the potential to improve accountability in police-civilian interactions in American public space, is evidently positive for both police and civilian. The evidence provided by the video enabled the police organization to show that Sousa used appropriate force consistent with department training and policy, and that Rodriguez did commit the felonies he was charged with (ninja636b, 2007; Wallman, 2007a, b; Citizens Police Review Board, 2007a, b, c; *State of Florida v. Rodriguez*, 2007).

The value of Hypothesis 2, that user-generated online video is significantly different from professional video journalism in its effects on accountability in police-civilian interactions in American public space, in this case does not appear to be evaluable without conjecture.

Hypothesis 3, that there are strong motivations for police to continue to attempt to restrict civilians' First Amendment rights to photograph police in public spaces, in this case appears to be negative; the video was material evidence in clearing Sousa (Wallman, 2007a, b).

Documents

Documents for this case include 15 data files, including one media clip from YouTube (ninja636b, 2007), one reader comment in a Yahoo forum (Walterlx, 2007), and the agenda and minutes of the Citizens Police Review Board meeting from the Fort Lauderdale city website (Citizens Police Review Board, 2007a, b, c). The two print media stories (Wallman, 2007a, b) were from ProQuest Newsstand, the original report from the local newspaper and the nearly identical wire service version. The story was not picked up by any other media, and was discussed in only the previously mentioned forum. The original story is no longer available on the newspaper's website. The Rodriguez criminal case summary data are available from the website of Broward County 17th Judicial Circuit of Florida (*State of Florida v. Rodriguez*, 2007). The case is listed as 'disposed'. A follow-up search on Frank Sousa was complicated by the fact that Sousa is the department spokesman, and has over 1,800 media citations. However, only the two versions (wire service and newspaper publication) of the single story mention both Sousa and either "kick" or Rodriguez.

Relevant Variables of Interest

The variable of **whether the police were aware of the camera** is evidently negative; throughout the video, no police appear to look toward the camera (ninja636b, 2007). The variable of **when the police became aware of the camera** is evidently not during the incident, according to the video. The variables of **whether there was any marked change in police behavior once they were aware of the camera**, **whether police made any attempt to prevent the recording**, **whether police made any attempt to acquire, confiscate, or destroy the video**, **whether the police gave any unlawful instruction to the videographer regarding the video**, and **whether police detained, cited, or arrested the videographer** are all negative (ninja636b, 2007).

The variable of **when the video was available via the Internet** is February 25, 2007, when the videographer, a German tourist with the first name Timo and the online nickname ninja636b uploaded the video to his YouTube channel (ninja636b, 2007). The variable of **when the video was available via broadcast news media** is apparently never, based on the document search and acquisition procedures, as is the variable of **when images from the video were available via print news media**.

The variable of **whether more than one camera captured the event** is positive. In the YouTube video, other camcorders and cell phone cameras are visibly in use (ninja636b, 2007). However, based on the document search and acquisition procedures, none of these other user-generated videos has become publicly available.

The variable of **whether official CCTV, dashboard, or other video captured the event** is negative. At the time of the incident, the Fort Lauderdale police did not have dashcams; the contract to install them was not signed until 2007 (Wallman, 2009). Because there were no police cameras on the scene, the variables of **whether police initially admitted to the existence or possession of video of the event**, **when did police admit to possession of video of the event**, **whether police released official video of the event**, and **whether official video of the event was available via the Internet** are also not applicable.

The variable of **who was credited as the source for each medium of release of the video** is only the ninja636b name on YouTube, because the video was not released in any other medium (ninja636b, 2007).

The variable of **whether there was any effort to restrict, remove or prosecute the release of the video** is apparently negative, based on the document search and acquisition procedures.

The variables of the **initial official response to handling of the videographer** and the **sequence of official actions regarding the videographer** are not applicable, because the police did not appear to be aware of the videographer (ninja636b, 2007).

The variables of the **final outcome regarding the videographer** and the **compensation of the videographer** are that the video gained a moderate number of views, but not enough to pay (ninja636b, 2007).

The variable of the **affiliation of the videographer** is unknown aside from his self-identification as being a German tourist (ninja636b, 2007).

The variable of **police misconduct** is negative. Sousa was cleared by Internal Affairs, the State Attorney's Office for Broward County, and by the Citizens Police Review Board (Wallman, 2007a, b; Citizens Police Review Board, 2007c). In addition, Rodriguez admitted that he was in the wrong and pled guilty to charges of disorderly intoxication, police battery, and resisting with violence (*State of Florida v. Rodriguez*, 2006).

The variable of the **initial response of the law enforcement agency regarding the officer involved** was to initiate an Internal Affairs investigation when it was informed of the YouTube video (Wallman, 2007a, b).

The variable of the **sequence of official actions regarding the officer involved** was: the department initiated an Internal Affairs investigation, including asking Rodriguez if he would like to file a complaint; he declined. At no time was Sousa suspended or otherwise removed from his police duties. The department's use-of-force trainer, Officer Carmelo Colon, reviewed the video and Sousa's report, and found that Sousa's application of a single kick to Rodriguez' brachial plexus was consistent with departmental policy and training. The investigation concluded with Internal Affairs Capt. Rick Maglione sending a memo to the police chief, who forwarded reports to the State Attorney's Office for Broward County, and to the Citizens Police Review Board. (Wallman, 2007a, b).

The variable of the **final outcome regarding the officer involved** was, the State Attorney's Office for Broward County issued a closeout memo on November 13, the Citizens Police Review Board approved the finding of "unfounded" on December 10, and Sousa was officially cleared (Wallman, 2007a, b; Citizens Police Review Board, 2007c).

For the variable of **final outcome of the law enforcement agency regarding policy changes, if any**: there were no changes to department policy (Wallman, 2007a, b).

The variable of the **affiliation of the civilian subject** appears to be entirely civilian, based on the document search and acquisition procedures. Rodriguez is not a journalist, nor is he affiliated with any media organization or with the police, or with any activist groups known for provoking the police (Wallman, 2007a, b).

The variable of **the initial official response to handling of the civilian subject** is that Sousa and Baro directed Rodriguez to leave the bar and nightclub district shortly after the bars closed at 4 a.m. Rodriguez, obviously intoxicated, became belligerent, shouting at the officers in Spanish (Wallman, 2007a, b).

The variable of the **sequence of official actions regarding the civilian subject** was: Sousa, who is fluent in Spanish, informed Rodriguez that the bar owners had previously filed trespass affidavits, which gave police legal authority to make arrests of persons who refused to leave. When Rodriguez continued to refuse to leave, Baro and Sousa attempted to arrest him. Only the last two minutes of this arrest are recorded in the YouTube video; as the video begins, Rodriguez is prone on the street between a taxi and the curb. Rodriguez visibly resists arrest, including putting his arms under his body so the police could not handcuff him. In order to gain control of Rodriguez' right arm, Sousa delivers one kick to the brachial plexus (00:16), a use of force that is taught to police and is approved by the department. Rodriguez continued to struggle, and at 00:27 a third officer joins the arrest. Rodriguez continued to protest in Spanish as the officers repeat, "Stop resisting!" At 00:52 a fourth officer joined the arrest; at 01:32 the fifth and sixth officers joined the arrest; at 01:50 Rodriguez is handcuffed, the officers stand up, and the video ends at 01:53 (ninja636b, 2007). Rodriguez was taken to jail, where he remained until a relative posted his bond (*State of Florida v. Rodriguez*, 2006).

Sousa documented his use of force in his report, including both his kick and his later application of pepper spray; he also stated that Rodriguez had punched him in the chest. Other police reported that Rodriguez spit at them. Felony charges filed September 4 against Rodriguez included battery on a law enforcement officer, resisting/obstructing an officer with violence, and disorderly intoxication (Wallman, 2007a, b).

The variable of the **compensation (if any) of the subject** is none; Rodriguez admitted that he was in the wrong and pled guilty to felony charges of disorderly intoxication, battery on a law enforcement officer, and resisting/obstructing an officer with violence (*State of Florida v. Rodriguez*, 2006).

The variable of the **final outcome regarding the civilian subjects** was that Rodriguez' case was disposed on September 21, 2006 (*State of Florida v. Rodriguez*, 2006).

The variable of **what third parties (if any) involved themselves in the legal case** is none, other than Rodriguez' attorney (*State of Florida v. Rodriguez*, 2006).

The variable of ***cui bono?*, or who benefits?**, in this case appears to be both the civilians and police of Fort Lauderdale. Evidently, the system worked, clearing an officer who did his job according to his training, and clarifying police actions as a measured and appropriate use of force to apprehend a violent drunk.

The variable of **how many times the video was viewed online** was over 1,700 by the time the investigation reports were released, and is over 16,000 as of February 2, 2012 (ninja636b, 2007). The variable of **how much press coverage the event received** included only the single story in the *Sun-Sentinel*, which was also distributed through a wire service (Wallman, 2007a, b). The variable of **how much Internet discussion was linked to this event** is a single post to a Yahoo forum, with a comment on the cultural differences between police-civilian interactions in the US versus those in Cuba, Rodriguez' country of origin (Walterlx, 2007).

CASE STUDY XIV: GRABER, UHLER ET AL., 2010-03-05

On March 5, 2010 civilian Anthony John Graber III videorecorded Maryland State Trooper First Class (Tfc) Joseph David Uhler making a traffic stop at the Interstate 95 exit to MD Route 543 in Harford County, MD. This case is included for direct replication (similar results for expected reasons), per Yin (2003, pp. 5, 110); the phenomenon is a police-civilian interaction in American public space during which police misconduct may have occurred, and of which user-generated video was posted online.

Research Question and Hypotheses

For Research Question 1, what is the outcome of user-generated online video on police-civilian interactions in American public space?, the outcomes of this case include: the criminal case against Graber was dismissed (*Maryland v. Graber*, 2010b); the state attorney general's office advised that recording public conversations between police and civilians is not a violation of the state's wiretap act (McDonald, 2010); the Maryland General Assembly briefly considered House Bill 45 to stop police from arresting civilians for recording them (Maryland General Assembly, 2011); Graber sold his motorcycle and filed for bankruptcy (Hermann, 2010; PACER, 2011), but avoided a felony conviction and kept his security clearance and his job.

For this case, Hypothesis 1, that user-generated online video has the potential to improve accountability in police-civilian interactions in American public space, appears to be supported. The court dismissed the false wiretap charges the police brought against the civilian.

The value of Hypothesis 2, that user-generated online video is significantly different from professional video journalism in its effects on accountability in police-civilian interactions in American public space, in this case does not appear to be evaluable without conjecture.

Hypothesis 3, that there are strong motivations for police to continue to attempt to restrict civilians' First Amendment rights to photograph police in public spaces, appears to be supported. The trooper was videorecorded violating procedure, and police actions in seeking to prosecute the videographer initiated the *Streisand Effect* (Masnick, 2005), drawing more attention to the video. Images and videos of the trooper have now been widely distributed on both the Internet and in traditional print and broadcast media. State authorities, including the legislature, have since acted to discourage similar police actions in the future, curbing police discretion.

Documents

The case data total more than 180 files, including eight media clips from YouTube and news media websites. There were 39 court documents; the original charge, the affidavits for warrants, Graber's motion to suppress, and the ruling for dismissal contain most of the important data (*Maryland v. Graber*, all dates). Other significant documents include the Application for Statement of Charges, an image of which was posted on a blog (Miller, 2010), and Graber's media interviews following the search of his home and his release from jail (McPherson, 2010; WMAR-TV, 2010a; Miller, 2010). There were over 90 documents from LexisNexis Academic news sources, and three law review articles. The Google search +Anthony +Graber +Maryland +police retrieves nearly 80 Google news articles (current plus archives), over 5000 blog posts, and over 2000 discussion forum posts. A Facebook wall supporting Graber has over 6000 likes (Facebook, 2011).

Relevant Variables of Interest

The variable of **whether the police were aware of the camera** is positive, as the police attested in the affidavit for the search warrant (*Maryland v. Graber*, 2010a, p. 2; Miller, 2010). The variable of **when the police became aware of the camera** is evident in the longer video at approximately 03:30; in the shorter video, it is almost immediate (Graber, 2010b, 03:30; 2010a, 00:02).

The variables of whether there was any marked change in police behavior once they were aware of the camera, whether police made any attempt to prevent the recording, whether police made any attempt to acquire, confiscate, or destroy the video, and whether the police gave any unlawful instruction to the videographer regarding the video are all evidently negative (Graber, 2010a, 2010b).

The variable of **whether police detained, cited, or arrested the videographer** is positive. The videographer was detained, Uhler issued him one citation for 80 mph in a 65 mph zone, then he was released. (*Maryland v. Graber*, 2010a, p. 2; Miller, 2010).

The variable of **when the video was available via the Internet** is March 10, 2010 for the short version, and March 12 for the long version (Graber, 2010a, 2010b). The variables of **when the video was available via broadcast news media** and **when images from the video were available via print news media** are April 9, 2010 (WMAR-TV, 2010a, 2010b; Taylor, 2010).

The variable of **whether more than one camera captured the event** is, based on the document search and acquisition procedures, that apparently only the single civilian camera captured the event (Graber, 2010a, 2010b). The variable of **whether official CCTV, dashboard, or other video captured the event** is apparently negative. Maryland State Police cruisers have mandatory dashboard cameras (*Maryland v. Graber*, 2010a, pp. 51-78; Shin, 2010), and the cruiser driven by Trooper Esh, visible when Graber looks over his shoulder (Graber, 2010a, 2010b), was in position to record relevant video. However, the court's ruling states that "no such recordings exist" (*Maryland v. Graber*, 2010b, p. 2). Thus, the variables of **whether police initially admitted to the existence or possession of video of the event**, **when did police admit to possession of video of the event**, **whether police released official video of the event**, and **whether official video of the event was available via the Internet** are all not applicable.

The variable of **who was credited as the source for each medium of release of the video** is that several television reports credited Graber verbally, while others credited YouTube or superimposed their own station logo. NPR credited nikotyc/YouTube for the embedded video on its website (Rose, 2011). Most print sources either omitted a credit entirely, or credited YouTube for still images taken from the video.

The variable of **whether there was any effort to restrict, remove or prosecute the release of the video** is positive; the prosecution of the videographer is the salient characteristic of this case (*Maryland v. Graber*, all dates).

The variables of the **initial official response to handling of the videographer** and of **the initial official response to handling of the civilian subject** are that the videographer was detained, Uhler issued him one citation for 80 mph in a 65 mph zone, then he was released (*Maryland v. Graber*, 2010a, p. 2; Miller, 2010).

On March 5, 2010 around 4:35 PM Uhler was traveling north on I-95 in an unmarked gray four-door sedan. He observed Graber driving his motorcycle at excessive speed, popping a wheelie, and cutting off at least one vehicle. Trooper Esh joined the pursuit, and Uhler established radio contact with Esh. Graber took the exit for MD Rt. 543; when he slowed as he approached vehicles stopped at the end of the exit, Uhler pulled diagonally in front of Graber's motorcycle while Esh pulled in behind the motorcycle (McGuire, 2010; Graber, 2010a, 2010b).

In the video, it is evident that Graber backs his motorcycle away from the sedan, which came within a few feet of his front tire. Uhler, in plain clothes and not displaying a badge or other police identification, exits his vehicle and immediately draws a semiautomatic pistol from his right hip, pointing it down and to his right, then says, "Get off the motorcycle. Get off the motorcycle!" Uhler then grabs the windscreen of the motorcycle, and continues, "Get off the motorcycle! State police" (Graber, 2010a, 2010b). Approximately five seconds elapse between Uhler's drawing his pistol and identifying himself as police. Graber later stated, "I was afraid. I thought the person, at the time I didn't know it was an officer, was going to shoot me or steal my bike" (WMAR-TV, 2010).

Graber complied with Uhler's directions. Uhler observed the camera mounted on Graber's helmet. Graber produced identification, and admitted to driving at excessive speed. Uhler issued him one citation for 80 mph in a 65 mph zone, then released him (*Maryland v. Graber*, 2010a, p. 2; Miller, 2010; McGuire, 2010). In his official report of the incident, Uhler omitted the fact that he had drawn his pistol (Miller, 2010).

The variables of the **final outcome regarding the videographer** and of the **final outcome regarding the civilian subject** are that Graber avoided a felony conviction and kept his government security clearance and his consulting job with a defense contractor, but sold his motorcycle at a loss, saying "I don't want to ever have a motorcycle again" (Hermann, 2010), and filed for bankruptcy (PACER, 2011).

The variables of the sequence of official actions regarding the videographer and of the sequence of official actions regarding the civilian subject are complex.

Detective Sergeant Mark McGuire reported that on March 15 he was "made aware of" a YouTube video of the incident (McGuire, 2010). Graber later said, "I posted it on YouTube because my mom was worried about the legality of it, and, she was upset that a police officer pulled a firearm on me because I'm not a criminal" (McPherson, 2010). Graber posted the short video on March 10, and the longer video without audio on March 12 (Graber, 2010a, 2010b).

On March 15, McGuire searched YouTube for the incident, and found Graber's videos on his nicotyc YouTube channel (McGuire, 2010). The longer video documented several traffic violations prior to Uhler's sighting of Graber, including a visible peak of 128 mph on the motorcycle's speedometer (Graber, 2010a, 2010b). McGuire viewed the videos with Uhler, who confirmed that the videos were consistent with what he had observed during the traffic stop (*Maryland v. Graber*, 2010a, p. 41). The next day, McGuire queried the Maryland Motor Vehicle Administration (MVA) for Graber, acquiring the home address of his parents, where he had recently moved. On March 18, McGuire spoke with an officer at Aberdeen Proving Ground, where Graber is in a reserve unit, and confirmed Graber's current home address. On March 22, McGuire ran a property search on the address, and found it listed to Graber's parents. The same day, he drove by that address, and noted the license plates of five vehicles, all of which plates he ran through MVA, confirming that two of the vehicles were registered to Graber (*Maryland v. Graber*, 2010a, pp. 41-42).

On April 7, McGuire requested a criminal warrant for Graber for violation of CR 10-402(a), the wiretap statute, and for additional traffic charges of reckless driving and of negligent driving. In his application for a search warrant, McGuire stated that he had been a state trooper for 14 years, was at the time in the Criminal Investigations Division, and that he believed "that the laws regulating intercepting any wire, oral, or electronic communications, were violated by Anthony John Graber III, using the video camera described above that was affixed to his helmet, and then subsequently downloaded to the internet" (*Maryland v.*

Graber, 2010a, pp. 42-43). The scope of the warrant was the property address, including all buildings, and to "seize all documentation, and any other evidence to include the above mentioned video recorder, photos, digital media, video, DVDs, CD's, storage devices, computers, or any other media found in or upon said residence" (*Maryland v. Graber*, 2010a, p. 43). It is worth noting that Graber was not the property owner, and that the household included his parents (the owners), his younger sister, his wife, and his two young children (Shin, 2010). The search warrant, as issued, encompassed the personal property of the entire family.

McGuire and five other troopers executed the search warrant at 6:45 AM the following morning, waking the family. When Graber pointed out that the search warrant had not been signed by a judge, "They told me they don't want you to know who the judge is because of privacy" (Miller, 2010). The entire family was detained for an hour and a half; Graber's mother was prevented from going to work, and his sister was prevented from going to school. The troopers seized two computers, two laptops, two external hard drives, a thumb drive, and the camera. Execution of the arrest warrant ran into a problem, because Graber had just had gall bladder surgery, and had bandages to show for it. After a call to headquarters, the troopers allowed Graber to remain at home, with the understanding that he would turn himself in when he had healed sufficiently (Miller, 2010; Shin, 2010).

Officials called a press conference later that day; news media had already interviewed Graber at his home (McPherson, 2010; WMAR-TV, 2010a). Greg Shipley, spokesman for the Maryland State Police, stated that the intent of charging Graber was to protect Maryland from illegal wiretapping: "We are enforcing the law, and we don't make any apologies for that" (McPherson, 2010). "He had been recording this trooper, audibly, without his consent. This information was taken to the State's Attorney's office in Harford County." The trooper drew his gun in the first place, according to Shipley, because Graber backed his bike up, "creating a brief moment of fear for the trooper" (The Aging Rebel, 2010). The trooper "held that gun at his side momentarily. When he saw the situation was under control he quickly put it away. Never pointed it at that individual. And we think he acted appropriately" (McPherson, 2010).

Officials initially set Graber's bond at $15,000 when he turned himself in on April 15, despite the fact that a wiretap conviction carries a maximum fine of only $10,000. He was held at Baltimore County Jail for 26 hours before being released on his own recognizance. According to Graber, "The judge who released me looked at the paperwork and said she didn't see where I violated the wiretapping law. She said, 'I have no idea why you're charged with this'" (Miller, 2010). Within hours of his release from jail, Graber was interviewed by Carlos Miller, the blogger of *Photography is not a crime* (now *Pixiq*). Shortly afterwards, Graber ceased making public statements on advice of counsel (Shin, 2010).

The prosecutor, State's Attorney for Harford County Joseph Cassilly, charged Graber with felony wiretapping under Maryland's two-party consent law (CNN, 2010). Cassilly and the state police made public statements that they were within the law (Shin, 2010).

On April 27, 2010, the Harford County Grand Jury indicted Graber on seven counts: three for reckless, negligent, and excessive speed driving, and the remainder for various aspects of wiretapping. If convicted, Graber could have been sentenced to up to 16 years in prison, and a conviction on any of the felony wiretap charges would have meant the loss of his security clearance and his job (ACLU-MD, 2010a; Hermann, 2010). On June 1, Graber was arraigned in Harford County Circuit Court (Maryland Judiciary, 2011).

On September 27, 2010, Judge Emory A. Plitt, Jr. dismissed all but the traffic counts of the indictment against Graber. Judge Plitt ruled that the video recording was not of a private conversation, and that since Graber's recording was conduct protected by the First Amendment, Maryland State law could not criminalize it (*Maryland v. Graber*, 2010b, p. 18).

The variables of the **compensation of the videographer** and of the **compensation (if any) of the civilian subject** are unknown, based on the document search and acquisition procedures.

The variables of the **affiliation of the videographer** and of the **affiliation of the civilian subject** appear to be entirely civilian. He is not a journalist, nor is he affiliated with any media organization or with the police, or with any activist groups known for provoking the police. However, the fact that he is a military reservist and a motorcycle enthusiast has had an effect on media exposure for this case, as evidenced by mentions in many of the press stories and blog and forum posts.

The variable of **police misconduct** is positive. Police violation of procedure – failure to identify as police – was recorded, and there was also some question of the appropriateness of drawing a pistol for a traffic stop (Graber, 2010a, 2010b).

The variables of the initial response of the law enforcement agency regarding the officers involved, the sequence of official actions regarding the officers involved, and the final outcome regarding the officers involved are essentially identical, because there has not been any official action against the police in this case. Official statements have consistently been that the troopers "acted appropriately" and that the investigation and prosecution was within the law (Shin, 2010; CNN, 2010; McPherson, 2010).

The variable of **final outcome of the law enforcement agency regarding policy changes, if any,** remains ambiguous. In July, the state attorney general's office advised that recording public conversations between police and civilians is not a violation of the state's wiretap act (McDonald, 2010). Cassilly announced in a subsequent radio interview he had no intention of abiding by the attorney general's advice (Balko, 2011). However, the court ruled,

> Those of us who are public officials and are entrusted with the power of the state are ultimately accountable to the public. When we exercise that power in a public forum, we should not expect our activity to be shielded from public scrutiny. (*Maryland v. Graber*, 2010b, p.)

Shipley reportedly said that the state police respect the judge's ruling, and that troopers would be informed of it. He also said that if they suspect a violation of the wiretap law, troopers are to present the case to prosecutors before filing charges (Nuckols, 2010).

In January, 2011, the Maryland General Assembly considered House Bill 45, summarized as: "This bill authorizes … a person to intercept an oral communication made by a law enforcement officer: (1) in a public place; and (2) in the course of the officer's regular duty" (Maryland General Assembly, 2011). The Judiciary Committee reported unfavorably, and the sponsor withdrew the bill on February 18 (Maryland General Assembly, 2011 February 18).

The variable of **what third parties (if any) involved themselves in the legal case** is that the ACLU of Maryland is a third party to the case. David Rocah, a staff attorney of the ACLU-MD, represented Graber, along with *pro bono* attorneys Joshua Treem, David Weinstein, and Nicholas Vitek from Schulman, Treem, Kaminkow & Gilden in Baltimore, and John Stewart, Ann Mace, and Cynthia Kendrick from Crowell & Moring in Washington, DC. (ACLU-MD, 2010b; Nuckols, 2010).

The variable of ***cui bono?*, or who benefits?**, in this case appears to be civilians in Maryland who choose to record police in public. If police follow the recommendations of the court and the attorney general, civilian videographers should no longer be arrested or charged under the wiretap statute.

The variable of **how many times the video was viewed online** is 1.2 million views on Graber's YouTube channel, plus 1.1 million views for the most popular duplicate, for a total in excess of 2.3 million views (Graber 2010a, 2010b; FunkensteinJr, 2010), aside from the many duplicate or mash-up videos that have garnered a few thousand views each.

The variable of **how much press coverage the event received** included, as of February 1, 2012, over 90 documents from LexisNexis Academic news sources, and the Google search +Anthony +Graber +Maryland +police retrieves nearly 80 Google News articles (current plus archives).

The variable of **how much Internet discussion was linked to this event** is that as of February 1, 2012, the Google search +Anthony +Graber +Maryland +police retrieves over 5000 blog posts, and over 2000 discussion forum posts. A Facebook wall supporting Graber has over 6000 likes (Facebook, 2011).

CHAPTER 5

Emergent Patterns and New Theories

Examination of the relevant variables of interest within and across cases has identified a number of common elements of the examined police-civilian interactions which are likely to recur in future similar interactions. These common elements present important implications for police accountability, municipal risk management, freedom of the press and other civil liberties, and broader theoretical issues that merit further study. At this point, it is useful to concisely review the methodology and goals of the present research in order to clarify the sources and purposes of the conclusions.

First, as used in the present research, a *theory* is an idea that clarifies an event or behavior. The phenomena examined in the present research are complex, deeply contextual, and often encompass very large quantities of data, thus presenting significant challenges to both the research analysis and to final understanding. Good theory is useful because it "…synthesizes the data, focuses our attention on what is crucial, and helps us ignore that which makes little difference" (Griffin, 1994, in Baran & Davis, 2006, p. 30). The theories presented here are the researcher's best descriptions of the phenomena, based on systematic analysis of the documents in each case.

Second, what the present research is *not*. It is not an attempt to present fully developed and tested explanatory theory; proving causality is beyond the goals of this *descriptive study*. The present research is also not a quantitative study; the fourteen cases represent separate qualitative investigations, not a population subject to statistical analysis.

There is not yet a sufficient body of data, including that produced by the present research, to make an explanatory case study feasible at this time. Thus, the methodology for the present research is based on *fourteen descriptive case studies*.

The phenomenon of a police-civilian interaction in American public space that has been videorecorded by a civilian presents appropriate material for a descriptive case study. A descriptive multiple-case study presents useful opportunities for both direct replication (similar results) and contrast (differing results for expected reasons) among cases, and for cross-case analysis (Yin, 2003b, 5, 110).

Following de Graaf & Huberts' (2008) solution, a 'monster grid' (p. 642) incorporates a row for each case and a column for each variable, thus presenting in a single document an at-a-glance summary of the 532 potential variables of interest. This document is presented in multiple-page format in the Appendix.

The goal of the present research was to generalize from the multiple-case study to useful theory (Yin, 2003a, 38). This theoretical framework "needs to state the conditions under which a particular phenomenon is likely to be found (a literal replication) as well as the conditions when it is not likely to be found (a theoretical replication)" (Yin, 2003a, pp. 47-48).

For the present research, analytical generalization to theory began with the grid. As de Graaf & Huberts (2008) summarize the process, "From this grid, patterns (in the form of propositions) were derived, which were then juxtaposed with the empirical data. This inductive process was repeated many times before the final analysis was written" (p. 642). Eisenhardt (1989) provides a more detailed description of the process:

> From the within-site analysis plus various cross-site tactics and overall impressions, tentative themes, concepts, and possibly even relationships emerge. The next step of this highly iterative process is to compare systematically the emergent frame with the evidence from each case in order to assess how well or poorly it fits with case data. The central idea is that researchers constantly compare theory with data - iterating toward a theory which closely fits the data. A close fit is important to building good theory because it takes

> advantage of the new insights possible from the data and yields an empirically valid theory. (p. 541)

Yin (2003a) cautions that "[analytical] generalization is not automatic, however. A theory must be tested by replicating the finding in a second or even a third [case], where the theory has specified that the same results should occur" (p. 37).

Multiple iterations of this analytical generalization process have identified the following common elements, which are proposed as theories. Each theory describes the conditions under which particular police-civilian interaction phenomena are likely to be found, and the conditions under which those same phenomena are not likely to be found (Yin, 2003a, pp. 47-48). It is hoped that these theories will merit further study, and that they may prove useful in clarifying the complex, richly contextual set of problems inherent to police-civilian interactions in American public space, particularly with the recent addition of user-generated online video to those interactions.

THEORY ONE

In police-civilian interactions where police destroy, falsify, fail to file, or omit data from required documentation, the existence of online video correlates with improved accountability as evidenced by police disciplinary actions. In such interactions without online video, police disciplinary actions are reduced or absent.

The analytical generalization of this theory is based on the primary documents (the online video and the police report of the police-civilian interaction), and on the variables of interest that contain evidence of police disciplinary actions, particularly including the variables of the initial response of the law enforcement agency regarding the officer involved, the sequence of official actions regarding the officer involved, and the final outcome regarding the officer involved.

For Case Study I, the positive value of the variable of when the video was available via the Internet indicates that this case matches the criterion of including an online video of the police-civilian interaction. According to the variable of the sequence of official actions regarding the officer involved, the trial board found the officer guilty of failing to issue a citizen contact receipt and to file a report. This case therefore matches the criterion of failing to file required documentation. The variable of the final outcome regarding the officer involved was that the officer was terminated by the police commissioner. This case therefore matches the criterion of police disciplinary action. Case Study I therefore directly replicates Theory One, producing similar results.

For Case Study II, the positive value of the variable of when the video was available via the Internet indicates that this case matches the criterion of including an online video of the police-civilian interaction. According to the variable of the sequence of official actions regarding the officer involved, the jury found the officer guilty of falsifying business records and offering a false instrument for filing. This case therefore matches the criterion of falsifying required documentation. The variable of the initial response of the law enforcement agency regarding the officers involved was that the officer was suspended for six months before he resigned. This case therefore matches the criterion of police disciplinary action. Case Study II therefore directly replicates Theory One, producing similar results.

Case Study III is not evaluable according to Theory One due to missing data; the official report (if one exists) of the police-civilian interaction is not publicly available.

For Case Study IV, the positive value of the variable of when the video was available via the Internet indicates that this case matches the criterion of including an online video of the police-civilian interaction. According to the variable of the sequence of official actions regarding the officer involved, the officer refused to provide a statement to the department. This case therefore matches the criterion of failing to file required documentation. Also according to the variable of the sequence of official actions regarding the officer involved, the department announced that the officer had been suspended pending investigation. This case therefore matches the criterion of police disciplinary action. Case Study IV therefore directly replicates Theory One, producing similar results.

Case Study V is not evaluable according to Theory One due to missing data; the two officers central to the case study were not permitted to file official reports. According to the variable of the initial response of the law enforcement agency regarding the officers involved, the department questioned all officers on the scene other than Pigott and Marchesona, and the Brooklyn District Attorney asked that neither Pigott nor Marchesona be interviewed by the NYPD. .

For Case Study VI, the positive value of the variable of when the video was available via the Internet indicates that this case matches the criterion of including an online video of the police-civilian interaction. According to the variable of the sequence of official actions regarding the officer involved, an external reviewer pointed out that statements from the officers involved were missing from the case file, and that the source of the civilian's injuries and his allegations of being previously struck by the police officer were never addressed. This case therefore matches the criteria of failing to file and of omitting data from required documentation. Also according to the variable of the sequence of official actions regarding the officer involved, the department announced that the officer had been suspended for 30 days without pay. This case therefore matches the criterion of police disciplinary action. Case Study VI therefore directly replicates Theory One, producing similar results.

For Case Study VII, the positive value of the variable of when the video was available via the Internet indicates that this case matches the criterion of including an online video of the police-civilian interaction. According to the variable of the sequence of official actions regarding the videographer, the officer issued a summons for a charge of disorderly conduct, which charge the prosecutor later found to be unsubstantiated. This case therefore matches the criterion of falsifying required documentation. According to the variable of the initial response of the law enforcement agency regarding the officers involved, the officer was suspended without pay. This case therefore matches the criterion of police disciplinary action. Case Study VII therefore directly replicates Theory One, producing similar results.

For Case Study VIII, the positive value of the variable of when the video was available via the Internet indicates that this case matches the criterion of including an online video of the police-civilian interaction. According to the variable of the sequence of official actions regarding the officers involved, two officers were disciplined for not having their dashboard cameras running. This case therefore matches the criteria of failing to file required documentation, and of police disciplinary action. Case Study VIII therefore directly replicates Theory One, producing similar results.

For Case Study IX, the positive value of the variable of when the video was available via the Internet indicates that this case matches the criterion of including an online video of the police-civilian interaction.

According to the variable of the sequence of official actions regarding the videographer, the officers initially charged the videographer with aiding the escape of a prisoner, which the prosecutor later found to be unsubstantiated. This case therefore matches the criterion of falsifying required documentation. According to the variable of the sequence of official actions regarding the officers involved, a department spokeswoman stated that the two police now face disciplinary actions. This case therefore matches the criterion of police disciplinary action. Case Study IX therefore directly replicates Theory One, producing similar results.

For Case Study X, the positive value of the variable of when the video was available via the Internet indicates that this case matches the criterion of including an online video of the police-civilian interaction. According to the variable of the initial response of the law enforcement agency regarding the officers involved, the police chief stated that the officers did not file a use of force report. This case therefore matches the criterion of failing to file required documentation. Also according to the variable of the initial response of the law enforcement agency regarding the officers involved, several police were suspended. This case therefore matches the criterion of police disciplinary action. Case Study X therefore directly replicates Theory One, producing similar results.

For Case Study XI, there is no online video of the police-civilian interaction. According to the variable of the initial official response to handling of the civilian subjects, the district attorney's office was unable to locate any record of the civilians' arrests. This case therefore matches the criterion of destroying or failing to file required documentation. Based on the document search and acquisition procedures, there are no public records for this case evidencing any police disciplinary actions. Case Study XI therefore theoretically replicates Theory One, producing contrasting results for predicted reasons.

For Case Study XII, there is no online video of the police-civilian interaction prior to the arrests being completed. According to the variable of the sequence of official actions regarding the civilian subjects, charging documents include statements by arresting officers that the civilians were engaged in gang activity, obstructing pedestrian traffic, obstructing vehicular traffic, and alarming and annoying pedestrians. Police were unable to produce evidence of the arrestees performing any of these actions, including physical evidence described in police statements, and multiple civilian eyewitnesses denied that any of the group were misbehaving in any way. This case therefore matches the criterion of falsifying required documentation. Based on the document search and acquisition procedures, there are no public records for this case evidencing any police disciplinary actions. Case Study XII therefore theoretically replicates Theory One, producing contrasting results for predicted reasons.

For Case Study XIII, the positive value of the variable of when the video was available via the Internet indicates that this case matches the criterion of including an online video of the police-civilian interaction. According to the variable of the sequence of official actions regarding the civilian subject, the police officer documented his use of force in his report, including both his kick and his later application of pepper spray. This case therefore fails to match any of the criteria of police destroying, falsifying, failing to file, or omitting data from required documentation. According to the variable of the sequence of official actions regarding the officer involved, at no time was Sousa). suspended or otherwise removed from his police duties. This case therefore matches the criterion of reduced or absent police disciplinary action. Case Study XIII therefore theoretically replicates Theory One, producing contrasting results for predicted reasons.

For Case Study XIV, the positive value of the variable of when the video was available via the Internet indicates that this case matches the criterion of including an online video of the police-civilian interaction. According to the variable of the initial official response to handling of the civilian subject, in his official report of the incident, the trooper omitted the fact that he had drawn his pistol. This case therefore matches the criteria of police omitting data from required documentation. Based on the document search and acquisition procedures, there are no public records for this case evidencing any police disciplinary actions. According to the variable of the sequence of official actions regarding the civilian subject, the state police stated that the trooper acted appropriately. This case therefore matches the criterion of reduced or absent police disciplinary action. Case Study XIV therefore does not replicate Theory One, producing contrasting results for unpredicted reasons.

In summary, Case Studies I, II, IV, VI, VII, VIII, IX, and X directly replicate Theory One. Case Studies XI, XII, and XIII theoretically replicate Theory One. Case Studies III and V are not evaluable for Theory One. One case study, XIV, does not appear to replicate Theory One.

Theory One would seem to suggest that online video may function as a counterbalance to police manipulation of official documentation. This possibility has significant implications for police accountability, particularly Fourth Amendment protections against search and seizure and arrest without probable cause. These implications increase the potential value of research that would either affirm or refute the empirical applicability of this theory.

Research methodologies that include quantitative analysis of existing documents appear to be particularly appropriate, because the number of variables to be evaluated is small, most of them are two-valued, and the data could be anonymized with little or no loss of validity. One significant challenge to such a study is the likelihood of a large disparity between the number of police-civilian interactions that do not include online video, and the relatively smaller number of interactions that do include online video.

THEORY TWO

In police-civilian interactions where a civil suit for police misconduct is settled successfully, the existence of online video of the interaction correlates with significantly higher settlements. In such interactions without online video, settlements are significantly reduced.

The analytical generalization of this theory is based on the existence of the online video, and on the variables of the compensation (if any) of the subject, the final outcome regarding the civilian subject, and the sequence of official actions regarding the civilian subject. The term "significant" in this theory represents a factor of four between the highest individual settlement of the lower, without-video group, and the lowest individual settlement of the higher, with-video group.

Case Study I is not evaluable for this theory because the civil suit was not settled successfully.

For Case Study II, the positive value of the variable of when the video was available via the Internet indicates that this case matches the criterion of including an online video of the police-civilian interaction. According to the variables of the compensation (if any) of the subject, the final outcome regarding the civilian subject, and the sequence of official actions regarding the civilian subject, the civilian subject received a settlement for $65,001 plus $25,000 in attorney's fees. This case therefore matches the criterion of successfully settling the civil suit. This settlement is significantly higher than the individual settlements in Case Studies XI or XII, nearly four times the $23,000 highest settlement among the 32 plaintiffs in those case studies. Case Study II therefore directly replicates Theory Two, producing similar results for expected reasons.

Case Study III is not evaluable for this theory because the civil suit data (if any) were not publicly available.

For Case Study IV, the positive value of the variable of when the video was available via the Internet indicates that this case matches the criterion of including an online video of the police-civilian interaction. According to the variables of the compensation (if any) of the subject, the final outcome regarding the civilian subject, and the sequence of official actions regarding the civilian subject, the civilian subject's estate received a settlement for $1.3 million, and the subject's daughter received a separate settlement for $1.5 million. This case therefore matches the criterion of successfully settling the civil suits. These settlements are significantly higher than the individual settlements in Case Studies XI or XII, 56 times and 65 times, respectively, the $23,000 highest settlement among the 32 plaintiffs in those case studies. Case Study IV therefore directly replicates Theory Two, producing similar results for expected reasons.

Case Study V is not evaluable for this theory because the civil suit was not settled successfully.

For Case Study VI, the positive value of the variable of when the video was available via the Internet indicates that this case matches the criterion of including an online video of the police-civilian interaction. According to the variables of the compensation (if any) of the subject, the final outcome regarding the civilian subject, and the sequence of official actions regarding the civilian subject, the civilian subject received a settlement of $150,000. This case therefore matches the criterion of successfully settling the civil suit. This settlement is significantly higher than the individual settlements in Case Studies XI or XII, over six times the $23,000 highest settlement among the 32 plaintiffs in those case studies. Case Study VI therefore directly replicates Theory Two, producing similar results for expected reasons.

For Case Study VII, the positive value of the variable of when the video was available via the Internet indicates that this case matches the criterion of including an online video of the police-civilian interaction. According to the variables of the compensation (if any) of the subject, the final outcome regarding the civilian subject, and the sequence of official actions regarding the civilian subject, the civilian subject received a settlement of $121,644.48, including attorney's fees. This case therefore matches the criterion of successfully settling the civil suit. This settlement is significantly higher than the individual settlements in Case Studies XI or XII, over five times the $23,000 highest settlement among the 32 plaintiffs in those case studies. Case Study VII therefore directly replicates Theory Two, producing similar results for expected reasons.

Case Study VIII is not evaluable for this theory because the civil suit was not settled successfully.

For Case Study IX, the positive value of the variable of when the video was available via the Internet indicates that this case matches the criterion of including an online video of the police-civilian interaction. According to the variables of the compensation (if any) of the videographer, the final outcome regarding the videographer, and the sequence of official actions regarding the videographer, the city agreed to pay the videographer a settlement of $170,000 for damages and attorney's fees. This case therefore matches the criterion of successfully settling the civil suit. This settlement is significantly higher than the individual settlements in Case Studies XI or XII, over five times the $23,000 highest settlement among the 32 plaintiffs in those case studies. Case Study IX therefore directly replicates Theory Two, producing similar results for expected reasons.

For Case Study X, the positive value of the variable of when the video was available via the Internet indicates that this case matches the criterion of including an online video of the police-civilian interaction. According to the variables of the compensation (if any) of the civilian subject, the final outcome regarding the civilian subject, and the sequence of official actions regarding the civilian subject, the county agreed to pay the civilian subject a settlement of $2 million. This case therefore matches the criterion of successfully settling the civil suit. This settlement is significantly higher than the individual settlements in Case Studies XI or XII, over 86 times the $23,000 highest settlement among the 32 plaintiffs in those case studies. Case Study X therefore directly replicates Theory Two, producing similar results for expected reasons.

For Case Study XI, there is no online video of the police-civilian interaction. According to the variables of the compensation (if any) of the subject, the final outcome regarding the civilian subject, and the sequence of official actions regarding the civilian subject, the two civilian subjects received settlements of $20,000 each, including attorney's fees. This case therefore matches the criterion of successfully settling the civil suit. This settlement is significantly lower than the individual settlements in Case Studies II, IV, or VII, less than a quarter of the value of the lowest settlement among the four plaintiffs in those case studies. Case Study XI therefore theoretically replicates Theory Two, producing different results for expected reasons.

For Case Study XII, there is no online video of the police-civilian interaction. According to the variables of the compensation (if any) of the subject, the final outcome regarding the civilian subject, and the sequence of official actions regarding the civilian subject, the 30 civilian subjects received settlements, including attorney's fees, of $23,000 for one plaintiff, $20,000 each for 15 plaintiffs, $9,000 for 13 plaintiffs, and $7,500 for one plaintiff. This case therefore matches the criterion of successfully settling the civil suits. These settlements are significantly lower than the individual settlements in Case Studies II, IV, or VII, less than a quarter of the value of the lowest settlement among the four plaintiffs in those case studies. Case Study XII therefore theoretically replicates Theory Two, producing different results for expected reasons.

Case Study XIII is not evaluable for this theory because there has been no civil suit.

Case Study XIV is not evaluable for this theory because there has been no civil suit.

In summary, six case studies with online videos directly replicate Theory Two, producing similar results for expected reasons. Two case studies without online videos theoretically replicate Theory Two, producing different results for expected reasons. Six case studies are not evaluable for Theory Two.

Notably, the civil rights suits in Case Studies VI, IX, and X were settled after this theory was submitted for review. The results indicate that this theory is accurately predictive.

Theory Two would seem to suggest that online video may correlate to greater police accountability, if one can measure accountability in terms of monies paid out by municipalities to settle police misconduct lawsuits. This theory may have value as an argument in persuading municipalities to carry out police reforms. This theory may be of particular value in considering such issues as the NYPD's stop-and-frisk tactics, which have been the subject of class action lawsuits and significant legal and government policy analysis in recent years (La Vigne et al., 2012; Zeidman, 2013; Harvard Law Review, 2013a; Lerman & Weaver, 2014). If the statistics presented in the cited studies are examined in light of potential settlement costs, Theory Two would seem to warn of the multiplication of financial risk as internetworked civilian cameras permeate the populations targeted by police stop-and-frisk tactics. Municipal risk management may therefore provide an additional argument in favor of stop-and-frisk policy change.

On the other side of the courtroom, Theory Two may inform the choices and strategies of plaintiff's attorneys. To both those ends, there appears to be value in research that would either affirm or refute the empirical applicability of this theory.

Research methodologies including quantitative analysis of existing documents appear to be particularly appropriate, because the number of variables to be evaluated is small, and most of the documents are public and readily searchable. As with Theory One, a significant challenge to such a study is the likelihood of a large disparity between the number of police-civilian interactions that do not include online video, and the relatively smaller number of interactions that do include online video. Additionally, it can be challenging to locate the final terms of settlements that are not entered into court records. Balancing those challenges is the appreciation of a wider audience for an empirically applicable theory that presents results in dollars and cents; opportunities for publication and presentation beyond scholarly venues are a distinct possibility.

THEORY THREE

In police-civilian interactions where a camera may have videorecorded police actions, exclusive police custody of that camera or its recording correlates with the video being lost, destroyed, reported as nonexistent, or concealed from the public. In such interactions where at least one copy of the video exists outside of police custody, the video more often remains intact and publicly available.

The analytical generalization of this theory is based on the variables of interest that contain evidence of at least one camera during the police-civilian interaction, the variables of interest that contain evidence of exclusive police custody of the video, the variables of interest that contain evidence of the loss, destruction, denial, or concealment of the video, and the variables of interest that contain evidence of the intact existence or publication of the video.

In more detail, these variables of interest include: whether more than one camera captured the event; whether official CCTV, dashboard, or other video captured the event; whether police made any attempt to acquire, confiscate, or destroy the video; whether police initially admitted to the existence or possession of video of the event; when did police admit to possession of video of the event; whether police released official video of the event; whether official video of the event was available via the Internet; when the video was available via the Internet; when the video was available via broadcast news media; and when images from the video were available via print news media.

The statement of theory uses the terms "video" and "camera" singularly; however, there are three case studies where multiple videos or cameras are analyzed individually.

For Case Study I, the positive value of the variable of when the video was available via the Internet indicates that this case matches the criterion of at least one camera during the police-civilian interaction, and the criterion of the intact existence or publication of the video. According to the value of the variable of how many times the video was viewed online, the video remained intact and publicly available beyond the end of the incident's legal sequelae. According to the variable of the sequence of official actions regarding the officer involved, the police did not obtain a copy of the video until after it was published online. This case therefore does not match the criterion of exclusive police custody of the camera or its recording, and matches the criterion for at least one copy of the video existing outside of police custody. Case Study I therefore theoretically replicates Theory Three, producing different results for predicted reasons.

For Case Study II, the positive value of the variable of when the video was available via the Internet indicates that this case matches the criterion of at least one camera during the police-civilian interaction, and the criterion of the intact existence or publication of the video. According to the value of the variable of how many times the video was viewed online, the video remained intact and publicly available beyond the end of the incident's legal sequelae. According to the variable of the initial response of the law enforcement agency regarding the officer involved, the police were not aware of the video until after it was published online. This case therefore does not match the criterion of exclusive police custody of the camera or its recording, and matches the criterion for at least one copy of the video existing outside of police custody. Case Study II therefore theoretically replicates Theory Three, producing different results for predicted reasons.

For Case Study III, the positive value of the variable of when the video was available via the Internet indicates that this case matches the criterion of at least one camera during the police-civilian interaction, and the criterion of the intact existence or publication of the video. According to the value of the variable of how many times the video was viewed online, the video remained intact and publicly available beyond the end of the incident's legal sequelae. According to the variable of the initial response of the law enforcement agency regarding the officer involved, the police were not aware of the video until after it was published online. This case therefore does not match the criterion of exclusive police custody of the camera or its recording, and matches the criterion for at least one copy of the video existing outside of police custody. Case Study III therefore theoretically replicates Theory Three, producing different results for predicted reasons.

Case Study IV presents several different data sets within a single case study because there were multiple videos, some of which passed through police custody and others which did not. Therefore, it is appropriate to test each video individually against Theory Three. For the Vargas video, the positive value of the variable of when the video was available via the Internet indicates that this case matches the criterion of at least one camera during the police-civilian interaction, and the criterion of the intact existence or publication of the video. According to the value of the variable of how many times the Vargas video was viewed online, the Vargas video remained intact and publicly available at least through February 2, 2012. According to the variable of whether police made any attempt to acquire, confiscate, or destroy the video, despite an attempt to confiscate Vargas' camera at the scene, the police did not obtain a copy of the Vargas video until after it was published online. The Vargas video therefore does not match the criterion of exclusive police custody of the camera or its recording, and matches the criterion for at least one copy of the video existing outside of police custody. The Vargas video of Case Study IV therefore theoretically replicates Theory Three, producing different results for predicted reasons.

Case Study IV also encompasses a video recorded by the CCTV system operated by BART. According to the variable of whether official CCTV, dashboard, or other video captured the event, the CCTV video matches the criterion of at least one camera during the police-civilian interaction. The same variable documents that the CCTV video was not released until the CD of the CCTV video was introduced at the preliminary hearing as People's Exhibit 27; this matches the criterion of exclusive police custody of the video. The variable of whether police initially admitted to the existence or possession of video of the event documents that for two days, BART officials maintained that the platform surveillance cameras did not record; this matches the criterion of the video being reported nonexistent. The CCTV video of Case Study IV therefore directly replicates Theory Three, producing similar results for predicted reasons.

Case Study IV also encompasses a cell phone camera used by Grant., as documented in the variable of whether the police were aware of the cameras; this matches the criterion of at least one camera during the police-civilian interaction. The same variable documents that Grant was able to use his cell phone camera to record an image of police shortly before he was shot, matching the criterion of a camera that may have videorecorded police actions. The variable of the initial official response to handling of the civilian subjects documents that, immediately following the shot, police began confiscating cameras, and that Grant's body was searched; this matches the criterion of exclusive police custody of the camera. Grant's recording was not delivered by police in response to subpoena, or recovered from Grant's camera; rather, the variable of whether there was any effort to restrict, remove or prosecute the release of the video documents that Grant's recording was available as courtroom evidence only because Grant was able to send it to the cell phone of his girlfriend before he was shot. This matches the criterion of the recording being lost, destroyed, reported as nonexistent, or concealed from the public. The Grant camera of Case

Study IV therefore directly replicates Theory Three, producing similar results for predicted reasons.

Case Study IV also encompasses a video recorded by Cross, as documented in the variable of whether police made any attempt to acquire, confiscate, or destroy the video; this matches the criterion of at least one camera during the police-civilian interaction. The same variable documents that police confiscated his camera's data card, which matches the criterion of exclusive police custody of the video. Based on the document search and acquisition procedures detailed in the methodology, there was no public release of the Cross video prior to its introduction as evidence in court. This matches the criterion of the recording being concealed from the public. The Cross video of Case Study IV therefore directly replicates Theory Three, producing similar results for predicted reasons.

For Case Study V, the positive value of the variable of when the video was available via the Internet indicates that this case matches the criterion of at least one camera during the police-civilian interaction, and the criterion of the intact existence or publication of the video. According to the value of the variable of how many times the video was viewed online, the video remained intact and publicly available beyond the end of the incident's legal sequelae. According to the variable of the initial response of the law enforcement agency regarding the officers involved, the department was apparently not aware of the video, and was still questioning officers on the scene when the video was published online. This case therefore does not match the criterion of exclusive police custody of the camera or its recording, and matches the criterion for at least one copy of the video existing outside of police custody. Case Study V therefore theoretically replicates Theory Three, producing different results for predicted reasons.

Like Case Study IV, Case Study VI encompasses multiple videos, some of which passed through police custody while one did not. For the Morris. video, the positive value of the variable of when the video was available via the Internet indicates that the Morris video matches the criterion of at least one camera during the police-civilian interaction, and the criterion of the intact existence or publication of the video. According to the value of the variable of how many times the video was viewed online, the Morris video remained intact and publicly available through February 2, 2012. According to the variable of the sequence of official actions regarding the videographer, the police did not obtain a copy of the Morris video until after it was published online. The Morris video therefore does not match the criterion of exclusive police custody of the camera or its recording, and matches the criterion for at least one copy of the video existing outside of police custody. The Morris video of Case Study VI therefore theoretically replicates Theory Three, producing different results for predicted reasons.

Case Study VI also encompasses videos of the incident recorded by police dashcams. According to the value of the variable of whether official CCTV, dashboard, or other video captured the event, the dashcam videos match the criterion of at least one camera during the police-civilian interaction. The same variable documents at least seven dashcam videos of the incident, only one of which was released to the DOJ investigators; the other six have remained unreleased, matching the criterion of exclusive police custody of the video. The values of the variables of whether police initially admitted to the existence or possession of video of the event, when did police admit to possession of video of the event, and whether police released official video of the event all match the criterion of the video being reported as nonexistent or being concealed from the public. The dashcam videos of Case Study VI therefore directly replicate Theory Three, producing similar results for predicted reasons.

For Case Study VII, the positive value of the variable of when the video was available via the Internet indicates that this case matches the criterion of at least one camera during the police-civilian interaction, and the criterion of the intact existence or publication of the video. According to the value of the variable of how many times the video was viewed online, the video remained intact and publicly available beyond the end of the incident's legal sequelae. According to the variable of the sequence of official actions regarding the videographer, the police did not obtain a copy of the video until after it was published online. This case therefore does not match the criterion of exclusive police custody of the camera or its recording, and matches the criterion for at least one copy of the video existing outside of police custody. Case Study VII therefore theoretically replicates Theory Three, producing different results for predicted reasons.

For Case Study VIII, the positive value of the variable of when the video was available via the Internet indicates that this case matches the criterion of at least one camera during the police-civilian interaction, and the criterion of the intact existence or publication of the video. According to the value of the variable of how many times the video was viewed online, the video remained intact and publicly available until the television station took down its YouTube channel. According to the variable of whether police made any attempt to acquire, confiscate, or destroy the video, the police did not evidently tamper with the camera, and did not appear to order either of the journalists to surrender or to destroy the recording. This case therefore does not match the criterion of exclusive police custody of the camera or its recording, and matches the criterion for at least one copy of the video existing outside of police custody. Case Study VIII therefore theoretically replicates Theory Three, producing different results for predicted reasons.

For Case Study IX, according to the variable of whether police made any attempt to acquire, confiscate, or destroy the video, police confiscated the camera, and while the cell phone was in police custody, videos of the incident were deleted. This case therefore matches the criteria of exclusive police custody of the camera or its recording, and of at least one camera during the police-civilian interaction, and does not match the criterion for at least one copy of the video existing outside of police custody. Case Study IX therefore directly replicates Theory Three, producing similar results for predicted reasons.

Like Case Studies IV and VI, Case Study X encompasses multiple videos, some of which passed through police custody while others did not. For the Winter video, the positive value of the variable of when the video was available via the Internet indicates that the Winter video matches the criterion of at least one camera during the police-civilian interaction, and the criterion of the intact existence or publication of the video. According to the value of the variable of how many times the video was viewed online, the Winter video has remained intact and publicly available beyond the end of the incident's legal sequelae. According to the variable of the initial response of the law enforcement agency regarding the officers involved, the police did not obtain a copy of the Winter video until after it was published online. The Winter video therefore does not match the criterion of exclusive police custody of the camera or its recording, and matches the criterion for at least one copy of the video existing outside of police custody. The Winter video of Case Study X therefore theoretically replicates Theory Three, producing different results for predicted reasons.

Case Study X also encompasses videos recorded by the CCTV system operated by the University of Maryland campus police, particularly the recording from Camera 158. According to the variable of when did police admit to possession of video of the event, the Camera 158 video matches the criterion of at least one camera during the police-civilian interaction. The same variable documents that the Camera 158 video was not released until subpoenaed, matching the criterion of exclusive police custody of the video. The same variable documents that the Camera 158 video was first reported missing, then when it was located, the critical time of the police-civilian interaction within the recording was found to have been damaged; this matches the criterion of the video being lost or being destroyed. The Camera 158 video of Case Study X therefore directly replicates Theory Three, producing similar results for predicted reasons.

Case Study XI is not evaluable for this theory because there were no cameras to record the police-civilian interactions.

Case Study XII is not evaluable for this theory because there were no cameras to record the police-civilian interactions.

For Case Study XIII, the positive value of the variable of when the video was available via the Internet indicates that this case matches the criterion of at least one camera during the police-civilian interaction, and the criterion of the intact existence or publication of the video. According to the value of the variable of how many times the video was viewed online, the video remained intact and publicly available beyond the end of the incident's legal sequelae. According to the variable of the initial response of the law enforcement agency regarding the officer involved, the police did not obtain a copy of the video until after it was published online. This case therefore does not match the criterion of exclusive police custody of the camera or its recording, and matches the criterion for at least one copy of the video existing outside of police custody. Case Study XIII therefore theoretically replicates Theory Three, producing different results for predicted reasons.

For Case Study XIV, the positive value of the variable of when the video was available via the Internet indicates that this case matches the criterion of at least one camera during the police-civilian interaction, and the criterion of the intact existence or publication of the video. According to the value of the variable of how many times the video was viewed online, the video remained intact and publicly available beyond the end of the incident's legal sequelae. According to the variable of the sequence of official actions regarding the videographer, the police did not obtain a copy of the video until after it was published online. This case therefore does not match the criterion of exclusive police custody of the camera or its recording, and matches the criterion for at least one copy of the video existing outside of police custody. Case Study XIV therefore theoretically replicates Theory Three, producing different results for predicted reasons.

In summary, case studies directly replicate Theory Three six times (IV-CCTV, IV-Grant, IV-Cross, VI-dashcam, X-Camera 158, XI), producing similar results for expected reasons. Case studies theoretically replicate Theory Three eleven times (I, II, III, IV-Vargas, V, VI-Morris, VII, VIII, X-Winter, XIII, XIV), producing different results for expected reasons. Two case studies, XI and XII, are not evaluable for Theory Three.

For the present research, Theory Three appears to be valid for multiple police jurisdictions and over the entire time span examined. Aside from the value of the content of each video to its respective case, the broader issues implicit in Theory Three are of significant public interest. Police denial of existence of official videos, police refusal to release videos without a court order, or the disappearance or erasure of videos while in police custody, may be indicators or symptoms of more serious problems, and may constitute research opportunities of significant value. There may be particularly rich opportunities for research on aspects of police transparency, based on the growing number of departments deploying dashcams and police-wearable cameras, and on the apparent wide range of policies and court rulings regarding access to the resulting recordings. Rachner's Seattle Police Video Project, cited in Case VI, appears to be worth investigation, documentation, and duplication.

There may also be opportunities for fruitful research in the outcomes of police-civilian interactions where a video recording of the interaction exists, but has never been made public. In this respect, Rachner's work in particular, and any similar projects regarding other departments, are likely to provide a wealth of research data in the form of official video. Research based on unpublicized user-generated video will be more challenging, as the video data will likely be limited to civil cases that were filed but which did not go to trial, and to internal investigations that also did not result in public trials. In either of these cases, access to the video data is likely to be challenging to secure. A third category, user-generated video not connected with any legal case or police investigation, and which has not been posted online, would probably require canvassing, advertising, or other means of identifying and contacting the videographers.

Video data for the aforementioned research is likely to become available from an increasing number of official sources. In-car video has become standard for many police organizations for a variety of reasons, and some of its effects have been examined in the literature (Klockars, et al., 2007). A logical extension of this technology is the wearable camera or Body-Mounted Video (BMV), such as the AXON system first tested by the San Jose police in 2009. The ear-mounted camera looks like a Bluetooth device, and records to flash memory (Baxter, 2009). Police organizations in Arizona, Florida, and Minnesota are also testing or have adopted the system since the first tests in California (Police in 3 States, 2011). At least one legal scholar has noted improvements in both police and civilian accountability correlated to the use of these devices, and proposes that these systems can improve police accountability regarding Fourth Amendment searches (Harris, D., 2009). At the time of this writing, no rigorous studies had yet been published, but at least one study-in-progress was reporting useful data (Stross, 2013) and the ACLU had published proposed guidelines (Stanley, 2013). The efficacy of these BMV systems presents a valuable research subject, of interest to municipal authorities, police, courts, civil rights advocates, and other stakeholders in police-civilian interactions.

APPENDIX

Relevant Variables of Interest

The following table contains the data from the 'monster grid' referenced in the text. The original working document was a horizontal scroll 11" in height and approximately 84" in width.

Case	Incident date	Police aware of camera	Became aware of camera	Marked change in behavior once aware of camera	Any attempt to prevent recording
I	7/1/07	YES	0:03:35	YES; redirects attention, asks if camera is on	YES; threatened videographer
II	7/25/08	NO	NO	NO	NO
III	9/5/09	YES	0:00:15	YES; moved to confront videographer	orders; interposing
IV	1/1/09	YES	0:00:00	NO; after shot, moved to confiscate cameras	YES; orders to stop, move away; interposing
V	9/23/08	YES	0:00:00	NO	Unknown
VI	4/17/10	YES	0:02:00	YES; treated detainee better	NO
VII	10/26/08	YES	0:00:00	NO	YES; ordered videographer to stop
VIII	4/15/05	YES	during stop	YES; ordered videographer to put down camera	YES; ordered videographer to put down camera
IX	10/1/07	YES	during incident arrest	NO	NO
X	3/3/10	YES	0:00:00	NO	NO
XI	8/20/09	N/A	N/A	N/A	N/A
XII	5/21/07	NO	After arrests	N/A	NO
XIII	9/3/06	NO	NO	NO	NO
XIV	3/5/10	YES	0:03:20	NO	NO

Case	Any attempt to acquire, confiscate or destroy video	Any unlawful instruction to videographer regarding video	Police misconduct was or may have been recorded
I	NO	YES; threat	YES; use of excessive force, discourtesy
II	NO	NO	YES
III	NO	YES; repeated; several	YES; unlawful instructions
IV	YES; moved to/did confiscate cameras	YES; ordered civilians to surrender cameras	YES; involuntary manslaughter
V	Unknown	Unknown	YES; violation of TASER procedure
VI	NO	NO	YES
VII	NO; only grabbed camera to detain videographer	YES; ordered videographer to stop	YES; false arrest, civil rights violation
VIII	NO	NO	YES; excessive force
IX	YES; arrested videographer, confiscated camera, deleted videos	YES	YES; excess force, unreasonable judgment
X	NO	NO	YES; excessive force; indictments
XI	N/A	N/A	YES; false arrest, civil rights violation
XII	NO	NO	YES; false arrests
XIII	NO	NO	NO; officer cleared
XIV	NO	NO	YES; violation of ID procedure

Case	Videographer was detained, cited or arrested	Initial response of law enforcement agency regarding the officer(s) involved
I	NO; temporarily detained with others	Miller attempted to file complaint; told supervisor was on vacation
II	NO	on desk duty as of July 28, pending investigation
III	NO; threatened	Company told guards not to repeat
IV	YES; detained to confiscate cameras	Mehserle taken to BART HQ; stated he thought Grant was going for a gun
V	Unknown	questioned on scene, then assigned desk duty pending investigation
VI	NO	internal affairs investigation begun day of incident; polce aware of YouTube video
VII	YES; detained, arrested	Suspension w/o pay pending investigation
VIII	YES; detained for ID	Officials refused to release information; after 30 days, WJLA filed internal complaint
IX	YES; detained, arrested	No action; IAD found nothing wrong
X	NO	Suspension pending investigation; no use of force report; "isolated incident"
XI	N/A	No action
XII	NO	No action; official statement of "Good-faith judgment"
XIII	NO	Internal Affairs investigation
XIV	YES; detained, cited	No action; statements that troopers acted appropriately

Case	Sequence of official actions regarding the officers involved
I	suspended with pay; investigated by IID; returned to duty; investigated by IID again; charged with misconduct; offered a plea deal; requested trial board; partially cleared by trial board; fired by commissioner
II	IAD investigation; DA indicted for assault and filing false paperwork; resigned before firing; jury verdict cleared assault, convicted on felony false paperwork
III	Withheld
IV	Mehserle suspended pending investigation; remained silent; resigned; fled to Nevada; charged with murder. BART board apologized to Grant family. Police chief retired under pressure. Domenici and Pirone fired.
V	statements that Pigott violated guidelines, was mistake, improper; criminal charges possible; city might not indemnify; investigation; contradictory statements to press following suicide
VI	Admin reassignment; internal investigation; criminal investigation; 30 day suspension w/o pay; demotion; retraining; return to duty; external reviewer noted gaps in case file
VII	Suspension without pay; return to duty
VIII	Officials say police followed procedure, acted appropriately, promised investigation; tapes reported missing; CCOP recommends discipline for not running cameras; internal affairs report sealed by court
IX	IAD reinvestigated after suit filed; court denied qualified immunity; Federal court ruled recording police has 1st Amend protection; IAD found "unreasonable judgment"; 2 officers disciplined
X	Internal, criminal and civil rights investigations; 4 suspended or on administrative leave; FBI/DOJ took over investigation; 2 indicted for assault and misconduct
XI	No action
XII	Brooklyn DA 'independent inquiry' confirmed police statements; no IAD investigation; no action
XIII	Internal Affairs; civilian subject refused to file complaint; use-of-force trainer reviewed video and report, stated consistent with approved policies and training; State Attorney, Citizens Police Review Board OK'd
XIV	No action; statements that troopers acted appropriately

Case	Final outcome regarding the officers involved	Effort to restrict, remove or prosecute release of video (takedown order, SLAPP, etc.)
I	Fired by police commissioner; appeal denied	NO
II	felony false report; barred from police; no jail time	NO; YouTube age restriction
III	Withheld	NO
IV	Mehserle convicted of involuntary manslaughter, 11 mo. jail; Domenici reinstated	NO
V	Pigott suicide; Marchesona returned to duty, promoted; suits dismissed, terminated	NO
VI	suspension w/o pay, demotion, reassignment, return to duty, overtime pay	YES; KCPQ refused to air, threatened legal action; videographer fired
VII	Returned to duty, assigned to public high school	NO
VIII	Ashton promoted to Sgt, transferred to Internal Affairs	NO
IX	IAD investigation complete; 2 officers disciplined	YES; deleted videos from cellphone camera; kept phone four months
X	McAleavey returned to duty; Baker cleared; Harrison convicted; Jones retired	NO
XI	No action	N/A
XII	No IAD investigation; no action; no admissions	NO
XIII	Cleared; followed procedures, filed complete and accurate reports	NO
XIV	No action; statements that troopers acted appropriately	YES; wiretap charges filed, warrant issued, property confiscated; arrest

Case	Final outcome of the law enforcement agency regarding policy changes, if any
I	complaint website; changed leadership, increased staffing, and mandated extra training for patrols in area; no release of officer names
II	no admission of guilt; no policy changes evident
III	Withheld
IV	Outside reviews; TASER retraining; tripled training hours; new chief; board meetings streamed live; 11-member Citizen Board selected
V	EDP retraining for ESU officers; new ESU commanding officer
VI	presumption of termination' for future incidents; OPA audit reports dashcam underuse; USDOJ finds police culture tolerates abuse; court-appointed monitor
VII	no admission of guilt; no policy changes evident; ACLU-NJ petitioned US DOJ to investigate the department
VIII	no admission of guilt; no policy changes evident; CCOP ongoing complaints about cameras not operating
IX	no admission of guilt; new training video based on incident; "no right of arrest for public and open recordings"
X	turnover in chiefs; politically significant
XI	no admission of guilt; no policy changes evident as yet
XII	no admission of guilt; no policy changes evident
XIII	None
XIV	Attorney General advised public recordings not wiretap offense; State's Attorney said would not abide; troopers informed of ruling; HB 45 considered, withdrawn

Case	Videographer affiliation	Initial official response to handling of videographer
I	civilian; high school; skateboarder	NONE
II	civilian; 3 kids in police dept.	NONE
III	civilian freelance photographer	Company statement
IV	civilians	Pirone prevented Grant from recording police; police attempted to/confiscated cameras
V	civilian	Unknown
VI	self-taught stringer	KCPQ editors contacted police re video, refused to air video
VII	journalist	Summons for disorderly conduct; council president talked to police, asked for investigation
VIII	journalist	Officials refused to release any information about incident
IX	civilian; attorney	arrested; phone confiscated; held until fee posted
X	civilian, student, filmmaker	N/A
XI	N/A	N/A
XII	journalist	N/A
XIII	civilian; German tourist	N/A
XIV	civilian; cycle fan; reservist	plainclothes trooper drew gun, slow to identify; detained, issued traffic citation; did not report gun

Case	Sequence of official actions regarding videographer	Videographer compensated
I	One interview with IID detectives	Unknown
II	called to testify during criminal trial	was paid $310 for video
III	Guards issued unlawful orders; company statement; company and agency refusal to release records	None
IV	interviewed by investigators; called to testify during preliminary hearings, trial	Unknown
V	Unknown	YES; amount withheld
VI	Police informed KCPQ of YouTube video; KCPQ fired videographer; IAD contact; KCPQ editors resigned, fired; subpoena for original media; gag order	paid $100 by KIRO-TV
VII	police arrest working news videographer on public sidewalk; summons for disorderly; unethical prosecution until change of venue; charges dismissed	YES; amount withheld
VIII	passenger in vehicle during felony stop; ordered to drop camera; frisked, ID'd, released; observed initiating officer smiling at them	not public information
IX	charged with wiretap, aiding escape, disorderly; charges dismissed on 1st Amend; IAD complaint filed; 1st IAD found no wrong	Suit settled for $170k
X	N/A	Unknown
XI	N/A	N/A
XII	N/A	Regular pay; not public
XIII	N/A	NO
XIV	Senior trooper informed of video, researches, gets warrant; seizes computers; civ. turns self in, released w/o bail; State's Atty prosecutes; case dismissed	Unknown

Case	Final outcome regarding videographer	Civilian subject affiliation	Compensation (if any) of the civilian subject
I	N/A	civilian; high school; skateboarder	Negative: civil suit dismissed, plaintiff to pay costs
II	inconvenience of testifying, but paid $310 for video	activist	Paid $65K plus att'y fees $25K, delayed to 12/10
III	Company statement of error; refusal to release records	Withheld	Withheld
IV	Grant shot, killed; no public record for other videographers	civilian, ex-con	YES; $1.5 mill. to daughter; $1.3 mill. to mother
V	Compensated; amount withheld; remains anonymous	psychiatric patient; EDP; HIV+, hep. C	None; suit from victim's mother incl. officer's estate; case terminated
VI	Still freelance; pled guilty to possession of stolen cameras	civilian	Civil suit settled; $150k
VII	Civil suit settled; no admission of guilt; payment delayed	journalist	Civil suit settled; $120k; payment delayed
VIII	Not injured; not party to civil suit	journalist	Jury trial awarded court costs plus $5000 medical expenses
IX	Suit settled $170k; cops not immune; findings letter critical	minor	Minor; data not released
X	N/A	civilian, student, family lawyers	Civil suit settled; $2m
XI	N/A	civilian	Paid $20K each, including att'y fees
XII	N/A	minors	Paid $9K-23K each; $447,500 total, incl. atty fees
XIII	No compensation; moderate number of YouTube views	civilian	NONE
XIV	No felony conviction; paid traffic fines; sold cycle; bankruptcy	civilian; cycle fan; reservist	Unknown

Case	Initial official response to handling of civilian subject
I	Warns skateboarders; returns to confront Bush; loses temper; excessive force; discourtesy; no report filed
II	police knocked civilian off bike, arrested, held for 27 hours, charged with misdemeanor assault with bicycle
III	Six trolley guards forcibly detained a man for smoking, a citable infraction
IV	Pirone used excessive force and profanity on compliant detainees, told Grant he could not take pictures; Mehserle reholstered his TASER, assisted Pirone in putting Grant facedown on the platform, drew his pistol, and shot Grant
V	investigation; placed Marchesona on modified duty; assigned Pigott to motor pool
VI	Police kicked, stomped innocent civilian, used ethnic slur in threat
VII	special' police officer forcibly arrested a news videographer, using excessive force, making threats, and violating civil rights; issued summons for disorderly conduct
VIII	county officials refused to release any information about the incident for over 30 days, denying WJLA's FOIA requests for 911 tapes and cell phone records of the police
IX	Minor; data not released
X	Beaten unconscious; arrested; coerced to conceal injuries; transported to jail
XI	unspecified criminal charges
XII	Detain, search, handcuff, and arrest 32 teens; charged with unlawful assembly, disorderly conduct
XIII	Instructed to leave premises per trespass affidavit; became belligerent
XIV	plainclothes trooper drew gun, slow to identify; detained, issued traffic citation; did not report gun

Case	Sequence of official actions regarding civilian subject
I	several interviews with IID detectives
II	dropped charges; settled civil rights suit; no admissions; city delayed paying
III	Withheld
IV	Initial statements blaming detainees; post-video claims that videos don't show everything; BART board apologized; repeated police perjury; settlement of multiple civil suits from victim's relatives
V	911 call requesting assistance; ESU requested; ESU attempted to talk subject down; subject hit unsecured officers with light tube; Pigott ordered tasing; Marchsona tased subject; no ESU officers broke fatal fall
VI	Public apology at press conference following video release; chief apology on return to city; subject files tort claim
VII	City officials made public statements supporting an investigation, but prosecutor and city attorney attempted to coerce a waiver by threatening prosecution. Charges dismissed after change of venue. Civil suit filed.
VIII	officials portrayed reporter as potential terrorist; police stalked reporter; department refused to cooperate with station; reneged on mediated settlement; ignored court orders; attempted sabotage, embarrassment
IX	Minor; data not released
X	charged with 2nd degree assault, disturbing the peace; dropped charges for lack of evidence
XI	arrest; transfer to booking; arraign; adjourn in contemplatin of dismissal (ACD); release on own recognizance; automatically dismiss charges for lack of evidence; settled civil rights suit before trial; no admissions
XII	Anonymous community board requests police; captain authorizes arrests; teens held at central booking, interrogated; offered community service; eventually dropped charges; settled civil rights suits; no admissions
XIII	After refusal to leave, arrested; Sousa placed one kick to brachial plexus; charged with Battery on LEO; Resisting w/violence; Disorderly Intox.; full reports filed by officers; jailed until bond; felony criminal prosecution
XIV	Senior trooper informed of video, researches, gets warrant; seizes computers; civ. turns self in, released w/o bail; State's Atty prosecutes; case dismissed

Case	Final outcome regarding civilian subject	Third parties (if any) involved in legal case	More than one camera captured event
I	Civil suit dismissed on technicality, plaintiff to pay costs	Mother filed civil suit; FOP Lodge #3 for Rivieri	NO
II	Paid $65K plus att'y fees $25K, delayed, no admissions or apology	Critical Mass	YES, other civilians
III	Withheld	Unknown	YES, at least one other civilian
IV	$1.5 mill. to daughter; $1.3 mill. to mother	NAACP; Amnesty Int'l; Meiklejohn; no amicus	YES, at least five civilians, CCTV
V	Civil suit from victim's mother, incl. officer's estate; case terminated	Pigott's widow sued city; case dismissed	YES, at least two other civilians
VI	Civil suit settled; $150k	NAACP, ACLU, El Centro de la Raza made statements; no amici briefs	NO
VII	Civil suit settled; $120k; payment delayed	ACLU of New Jersey	YES, at least one other civilian
VIII	Jury verdict excessive force, no civil rights violation; no apology	RCFP made statements, but no amicus brief	NO
IX	Minor; data not released	ACLU; BPPA; amicus briefs: CCR, Copwatch, CUAPB, Justice Committee, MPAC, NKCD	NO
X	Civil suit settled; $2m	NONE	YES, other civilians
XI	Paid $20K each, including att'y fees	NONE	N/A
XII	Paid $9K-23K each; $447,500 total, incl. atty fees	NONE	YES, at least one civilian, and news helicopter
XIII	Pled guilty; case disposed	NONE	YES, at least one other civilian
XIV	No felony conviction; paid traffic fines; sold cycle; bankruptcy	ACLU Maryland	NO

Case	*Cui bono*?	Video available via Internet	Video available via broadcast	Images from video available via print media
I	Department reputation for transparency; Baltimore citizens; skateboarders get park?	2/9/08	2/9/08 23:00	2/12/08
II	Civilian subject, attorney; cyclists	7/28/08	7/28/08 19:24	7/28/08 0:00
III	Heritage Security Services; videographers? Transit riders?	9/7/09	9/18/09 23:00	Never
IV	Plaintiffs' attorney; family; BART riders, reformers; Domenici got back pay	1/4/09	1/3/09	1/7/09
V	Civilians, esp. EDPs, re: TASER use	09/24/08	9/24/08 19:00	09/25/08
VI	Civilian subject; civilians in Seattle jurisdiction	4/20/2010	5/6/10 23:00	5/6/2010
VII	Videographer, attorney; no community benefits evident	10/26/2008	10/26/08 18:33	10/27/2008
VIII	PGC PD, county officials not made accountable; reporter vindicated when Johnson indicted	5/13/2005	5/13/05 17:00	5/13/2005
IX	videographers; 1st, 4th Amendment users	2/1/2010	12/22/2010	Not evident; no earlier than 2/1/2010
X	Civilians and their attorneys	3/4/2010	4/12/2010	4/13/2010
XI	Civilians and their attorneys	N/A	N/A	N/A
XII	30 civilians and their attorneys	4/21/2009	5/21/2007	NO
XIII	Police and civilians of Fort Lauderdale; the system worked	2/25/2007	NO	NO
XIV	Civilians in Maryland who record police in public	3/10/2010	4/9/2010	4/9/2010

Case	Official CCTV, dash-board, or other video captured the event	Police initially admitted to the existence or poss-ession of video of event	When did police admit to possession of video of event
I	NO	NO	N/A
II	NO	N/A	N/A
III	N/A	N/A	N/A
IV	YES, BART station CCTV	NO; camera was 'not working'	1/3/09; camera 'recorded but did not show incident'
V	NO	N/A	N/A
VI	YES, per Rachner	NO	Not as of 2/2/12
VII	NO	N/A	N/A
VIII	PGC PD denied cameras were working	N/A	N/A
IX	NO	N/A	N/A
X	Disputed; Camera 158 recording damaged	NO; denied existence	Discovery phase of pretrial investigation
XI	NO	NO	NO
XII	Unknown	N/A	N/A
XIII	N/A	N/A	N/A
XIV	NO	N/A	N/A

Case	Police released official video of event	Official video of event was available via Internet	For each medium of release of the video, source credited as
I	N/A	N/A	camb0i; YouTube; station ID
II	N/A	N/A	anonymous tourist; YouTube
III	N/A	N/A	Rob Hurlbut; theworldisraw
IV	06/04/09 preliminary hearing	YES; after introduction at trial	Varied; station ID, YouTube, videographer name
V	N/A	N/A	amateur video'; New York Post; Racquel McDonald
VI	NO	NO	Jud Morris; YouTube; KIRO-TV; 'freelance cameraman'
VII	N/A	N/A	WCBS-2, YouTube
VIII	N/A	N/A	WJLA ABC 7
IX	N/A	N/A	Glik; ACLU; YouTube; Greater Boston
X	Missing/ erased; subpoenaed	NO	YouTube; WJLA; AP; Winter
XI	NO	NO	N/A
XII	N/A	N/A	Cropped out
XIII	N/A	N/A	ninja636b
XIV	N/A	N/A	YouTube; Anthony Graber; nikotyc/YouTube

Case	Times video viewed online	Internet discussion linked to the event	How much press coverage the event received
I	5.5 million +; several 1 million +; 10 million total	78,000 comments on YouTube; 800 blogs; 1500 forums	All Baltimore; GMA; networks; wire services; national, regional, local papers
II	2.8 million+	10,000 comments on YouTube; 1700 blogs; 1500 forums; 240 videos	National; all NYC; networks; wire services; national, regional, local papers
III	18,879	YouTube; Pixiq; San Diego Reader; The World is Raw	9/18/09 NBC 7/39 San Diego ; 2/24/2010 San Diego Reader; no other press coverage
IV	12 mill.+ for 2274 YouTube videos	20k+ comments on top 3 videos; 52k blogs; 18k forums; 4500 videos	1000s of reports; national; int'l; LA; Bay Area; networks; wire services; subaltern
V	117K+	450 YouTube comments; Wikipedia; Facebook; 681 unique Google hits	National; all NYC; networks; wire services; tabloid press
VI	2500 unedited; not less than 300k+ aggregate	3000+ YouTube comments; 1400 blogs; 400 forums; 400 videos	International; national; regional; wire services; 61 unique Google News hits
VII	YouTube 29k; WCBS withheld; 78k+ total	40 forums; 18 blogs; 4 videos; several hundred video comments	CBS; AP wire service; local papers
VIII	Not publicly available	200 forums; 4k blog entries	ABC network; nat'l newspapers; press releases; wire services; industry journal; editorials
IX	15k+	50 video comments; 8500 blog posts; 3400 forum posts; 200 videos	National; Internat'l; Boston; networks; wire services; law reviews; 117 Google News hits
X	900K+ on YouTube	4k YouTube comments; 2.3k videos; 18k blog posts; 3k forum posts	119+ stories; National; Int'l; regional; networks; wire services; GMA; SI; ESPN
XI	N/A	Low single digits; comments on blog, newspaper websites	Single reports in NYC newspapers NYDN, Village Voice; Gothamist blog
XII	N/A	355 blog posts; 200+ website comments	2 TV, 3 radio reports; 12 print reports after arrests; 18 print reports after 1 year; 7 post-settle
XIII	16,161	1 Yahoo forum post, commenting on Cuban cultural difference	Single report in local paper & website, picked up by wire service
XIV	2.3 mill.+ for several videos	5k+ blog posts; 2k+ forum posts; Facebook wall 6k+ likes	90+ reports LexisNexis Academic; 80 Google News articles

References

... and uncovering lies. (2007, June 14). *New York Daily News*. Editorial. New York, NY. Retrieved from http://www.nydailynews.com/opinion/uncovering-lies-article-1.220130.

ABC News (ABCNews). (2010, April 14). Student beaten by cops fights back. *YouTube*. Retrieved from http://www.youtube.com/watch?v=b-RpmUN9irY.

ABC-7. (2005, May 12). ABC-7 files complaint against PG police for excessive force against reporter. Press release. *U.S. Newswire*. Retrieved from ProQuest Newsstand.

Abramson, A. (1987). *The History of Television, 1880 to 1941*. Jefferson, N.C.: McFarland.

Abramson, A. (2003). *The History of Television, 1942 to 2000*. Jefferson, NC: McFarland.

ACLU. (2010, February 1). Attorney challenges own arrest for recording police use of force in public. Press release. Retrieved from http://www.aclu.org/racial-justice/attorney-challenges-own-arrest-recording-police-use-force-public.

ACLUMASS1. (2011, June 2). The courage to get involved. *YouTube*. Retrieved from http://www.youtube.com/watch?v=lhd5_6DHV5s.

ACLU-NJ. (2010). In the matter of a petition for an investigation into the Newark, New Jersey Police Department by the United States Department of Justice pursuant to 42 U.S.C. § 14141. American Civil Liberties Union of New Jersey. Retrieved from http://www. aclu-nj.org/downloads/090910NPDUSDOJPEtition.pdf.

ACLU-NJ. (2011). Day 11: Quodomine v. City of Newark. *The Settlement Project: Citizen Edition*. American Civil Liberties Union of New Jersey.

Retrieved from http://www.aclu-nj.org/theissues/policepractices/ thesettlementprojectcitize/ day11quodominevcityofnewar/.

Adams, W.I.L., Ed. (1887, January 7). Photographic Pastimes. *The Photographic Times and American Photographer. 17*(277), 10-11. Scovill Manufacturing Company.

Aifantis, K. E., Hackney, S. A., & Kumar, R. V. (2010). *High Energy Density Lithium Batteries: Materials, Engineering, Applications*. Weinheim: Wiley-VCH.

ajajaj1. (2009, January 6). BART shooting - Updated!!! New angle of the shooting!!!! *YouTube*. Retrieved from http://www.youtube.com/watch?v=ZKKQ-gzc_Yw.

Albee, S.V. (1877). No. 32: Opposite 13th street, looking down the track, switches connecting with Pan-Handle R. R. In *The Railroad War at Pittsburgh, July 21-22, 1877*. [albumen stereograph]. Second Century Acquisition Fund (95.17.32). Carnegie Museum of Art, Pittsburgh, PA. Retrieved from: http://www.cmoa.org/searchcollections/details.aspx? item=1035643.

Alcatel-Lucent. (2011). Bell Labs Top 10 Innovations. Retrieved from http://www.alcatel-lucent.com/wps/portal/BellLabs/History.

Alexander, A. (2007, October 3). Bushwick 32: An example of racial profiling. New York: WBAI. Retrieved from http://archive.wbai.org/files/mp3/071003_080001wuc.MP3.

AmateursGuide. (2008a, February 12). #67 We know this cop. BPD Officer suspended after video is posted on Youtube. *Baltimore Sun talk forum*. Retrieved from http://talk.baltimoresun.com/showthread.php?p=3031359&page=5.

AmateursGuide. (2008b, February 13). :mad: How suspended officer also harassed an respectful professional. *Baltimore Sun talk forum*. Retrieved from http://talk.baltimoresun.com/showthread.php?t=135268.

American Civil Liberties Union of Maryland. (2010a, May 28). Backgrounder on Graber case. *ACLU-MD.org*. Retrieved from http://www.aclu-md.org/aPress/Press2010/ Graber_Factsheet.pdf.

American Civil Liberties Union of Maryland. (2010b, September 27). Wrongful charges dismissed against motorcyclist prosecuted for videotaping encounter with police. Press release. Retrieved from http://www.aclu-md.org/aPress/Press2010/092710_Graber.html.

American Civil Liberties Union, & White, E. J. (2011, January 13). *The ACLU & Elon James White: What to do when you're stopped by police - YouTube*. Retrieved from http://www.youtube.com/watch?v=ti2-NjnalFU.

American Civil Liberties Union. (n.d.a). Know your rights when encountering law enforcement. *American Civil Liberties Union.* Retrieved from http://www.aclu.org/national-security/know-your-rights-when-encountering-law-enforcement.

American Civil Liberties Union. (n.d.b). Know your rights when encountering law enforcement. Retrieved from http://www.aclu.org/files/kyr/kyr_english.pdf.

American Law Institute. (1977). *Restatement of the Law, Second, Torts.* Philadelphia: The Institute.

AmericanFascism. (2009, September 28). New Jersey Cop Assaults TV Cameraman. *YouTube*. Retrieved from http://www.youtube.com/watch?v=f5ixF_xHqYM.

Anonymous. (2008a, February 14). A very candid camera. To the point (Editorial). *Baltimore Sun.*

Anonymous. (2008b, February 18). Officer owes apology to skater and public. New ways for public to play political role. (Editorial). *Baltimore Sun.*

Answers Corp. (2011). Smartphones – truly personal. *Hot Topics and Trends.* Retrieved from http://www.answers.com/topic/hot-topics-and-trends.

Arendt, H. (1958). *The Human Condition.* Chicago, IL: University of Chicago Press.

Aristophanes. (ca. 393 B.C.E.). The Ecclesiazusae. In *Aristophanous Kōmōidiai* (Trans. Benjamin Bickley Rogers). London: G. Bell & Sons, 1902.

Arlen, M. J. (1979, April 23). The air: the tyranny of the visual. *The New Yorker*, 125.

Arlen, M. J. (1981). The tyranny of the visual. In *The Camera Age: Essays on Television* (pp. 185-198). New York, NY: Farrar Straus Giroux.

Arner, M. (2009, July 21). Trolley crimes could be linked. *Sign On San Diego.* Retrieved from http://www.signonsandiego.com/news/2009/jul/21/1m21trolley003633-trolley-crimes-could-be-linked/.

Arrest of a hobo. (ca. 1890). [gelatin silver print]. (X2010.11.2624, MNY216895). Museum of the City of New York. 1220 Fifth Avenue, New York, NY. Retrieved from http://collections.mcny.org.

Arthur, C. (2011, June 5). How the smartphone is killing the PC. *The Guardian.* Retrieved from http://www.guardian.co.uk/technology/2011/jun/05/smartphones-killing-pc/print.

Associated Press. (2005, May 17). Reporter claims mistreatment during Maryland traffic stop. Retrieved from http://www.firstamendmentcenter.org/reporter-claims-mistreatment-during-maryland-traffic-stop.

Associated Press. (2010a, April 13). Video allegedly shows Md. police beating student. *YouTube*. Retrieved from http://www.youtube.com/watch?v=TAPwyodTkYA.

Associated Press. (2010b, April 14). Md. police: second suspension in student beating. *YouTube*. Retrieved from http://www.youtube.com/watch?v=nNSxorQXWyw.

Associated Press. (2012, October 19). Jury finds 1 officer guilty, 1 innocent in 2010 beating of University of Md. student. *FoxNews.com*. Retrieved from http://www.foxnews.com/us/ 2012/10/19/jury-finds-prince-george-officer-guilty-assault-and-2nd-officer-cleared/

Astruc, A. (1948, March 30). Naissance d'une nouvelle avant-garde. *L'Ecran Français*, 144. transl. in *The New Wave: Critical Landmarks*, by Peter Graham, ed. New York: Doubleday, 1968.

Augenstein, N. (2009, September 22). Skateboarder's suit against Baltimore cop tossed. Baltimore, MD: WTOP Radio Network. Retrieved from http://www.wtop.com/ ?nid=858&sid=1768338.

Babwin, D. (2010, April 17). Cops getting caught on camera increases scrutiny. *New York Times*. New York. Retrieved from http://www.nytimes.com/aponline/2010/04/17/us/ AP-US-Cops-on-Camera.html.

Bain, G. G. (1908). Anarchist bombing, Union Square, New York City, March 1908. Bain Collection, Lot 10796. Washington, DC: Library of Congress. Retrieved from http://www.loc.gov/pictures/item/ 2002696063/.

Bain, G. G. (1909). Chinatown, N.Y.C. - police and detectives guarding Chinatown, July 6, 1909. Bain Collection, Lot 7178. Washington, DC: Library of Congress. Retrieved from http://www.loc.gov/pictures/item/2001704348/.

Balko, R. (2011, January 1). The war on cameras: it has never been easier--or more dangerous--to record the police. *Reason, (42)*8, 22. Retrieved from LexisNexis Academic.

Balko, R. (2011, January). The war on cameras: It has never been easier—or more dangerous—to record the police. *Reason*. Retrieved from http://reason.com/archives/2010/12/07/the-war-on-cameras/singlepage.

Baltimore City Police. (2008, May 1). How do I file a complaint about an officer? [web page] Retrieved from:

http://www.baltimorepolice.org/contact-us/faqs/136-contacting-the-bpd-for-non-emergencies/264-how-do-i-file-a-complaint-about-an-officer.

Baran, S. J. & Davis, D. K. (2006). *Mass Communication Theory: Foundations, Ferment, and Future*. 4th ed. Belmont, CA: Thomson Wadsworth. pp. 29-37

Barish, S. (2008, September 24). Naked man falls to death after being Tasered by cops. *CW11.com*. Retrieved from wpix.com/wpix-naked-man-falls-to-death-after-67633,0,7890871.story.

Barkan, S. E., & Bryjak, G. J. (2011). *Fundamentals of Criminal Justice: A Sociological View*. (2nd ed.). Sudbury, MA: Jones & Bartlett Learning.

Barker, C. J. (2007). Courtroom melee in Brooklyn. *New York Amsterdam News, 98*(50), 30. Retrieved from EBSCOhost.

Barker, C. J. (2008a). Bushwick 32 case back in court. *New York Amsterdam News, 99*(3), 3. Retrieved from EBSCOhost.

Barker, C. J. (2008b). Bushwick 32 teacher back in court. *New York Amsterdam News, 99*(6), 3. Retrieved from EBSCOhost.

Barker, C. J. (2009a). City to pay more than $250,000 to "Bushwick 32"; cases dismissed. *New York Amsterdam News, 100*(17), 3. Retrieved from EBSCOhost.

Barker, C. J. (2009b). Attorney Warren fights for "Brooklyn Four". *New York Amsterdam News, 100*(26), 32. Retrieved from EBSCOhost.

Barnouw, E. (1990). *Tube of Plenty: The Evolution of American Television* (2nd ed.). New York: Oxford University Press.

Barr, H. J. (1999). *The Jump-Out Boys*. New York, NY: Vantage Press.

BART. (2011, July 11). Public assured of independent oversight of officer-involved shooting from first-ever police auditor. Retrieved from http://www.bart.gov/news/articles/2011/ news20110711.aspx.

Bassett, R. K. (2007). To the Digital Age: Research Labs, Start-Up Companies, and the Rise of MOS Technology. Johns Hopkins University Press.

Baxter, S. (2009, November 2). San Jose Police will be first to use ear-mounted video cameras to record arrests. *The Mercury News*. San Jose, CA.

bbbbill123. (2008, February 13). Oreilly on Baltimore Cop. Retrieved from http://www.youtube.com/watch?v=fJvAZYaqPKI.

Bensinger, C. (1980). *The Video Guide* (Second). Berkeley, CA: Bookpeople.

Benson, T. W. & Anderson, C. (1989). *Reality Fictions: the Films of Frederick Wiseman*. Carbondale: Southern Illinois University Press.

Bentley, C. S. (2008). The Great Information Equalizer. *New American, 24*(8), 41-42.

Bezner, L. C. (1999). *Photography and Politics in America: From the New Deal into the Cold War*. Baltimore, MD: J. Hopkins University Press.

Bittner, E. (1970). The Functions of the Police in Modern Society: A Review of Background Factors, Current Practices, and Possible Role Models. Chevy Chase, MD: National Institute of Mental Health, Center for Studies of Crime and Delinquency.

Black, A. D. (1968). *The People and the Police*. New York: McGraw-Hill.

Blanchard, D. (2009, May 18-20). Hearing notes from inside the courtroom. Retrieved from http://www.indybay.org/newsitems/2009/05/22/18597048.php.

Booz & Company. (2011). Kodak to improve profit margin for digital cameras. *Management Consulting Case Interviews*. Consulting Case101. Retrieved from http://www.consultingcase101.com/kodak-to-improve-profit-margin-for-digital-cameras/.

Bouch, J. (2009, March). A picture is worth a thousand words. *Advances in Psychiatric Treatment, 15*(2), 81. Retrieved from http://apt.rcpsych.org/cgi/content/full/15/2/ 81.

Bowens, T. (2007, December 9). "Bushwick 32" supporters arrested on phony charges. *Workers World*. Retrieved from http://www.workers.org/2007/us/bushwick_32-1213/.

BoxCameras.com. (2008). *The Eastman (Kodak) Company No. 2 Kodak Camera, 1889-1897*. Retrieved from http://www.boxcameras.com/no2kodak.html.

Boyle, D. (1992). From Portapak to camcorder: A brief history of guerrilla television. *Journal of Film & Video, 44*, 67-79.

Bracklow, R. L. (1891). *Maiden Lane*. [gelatin silver print, 10 in. w. 8 in. h.]. (93.91.173, MNY206). Museum of the City of New York. 1220 Fifth Avenue, New York, NY. Retrieved from http://collections.mcny.org.

Brady, M. (ca. 1844). Superintendent of Metropolitan Police John A. Kennedy. Daguerreotype photograph. Retrieved from http://www.loc.gov/pictures/item/2004664011/.

Brandon v. Allen, 645 F. Supp. 1261, 1266-67 (W.D. Tenn. 1986).

Brandus, P. (2005, May 13). ABC-7 files complaint against police for excessive force. WTOP.com. Retrieved from http://www.wtop.com/?nid=&sid=518511.

Brendan, K. (2008, December 11). 'YouTube' skateboarder can sue cop, rules Baltimore City Circuit judge. *Daily Record, The* (Baltimore, MD), Retrieved from EBSCOhost.

Brendan, K. (2009, September 22). Skateboarder loses suit against Baltimore police officer. *Daily Record, The* (Baltimore, MD), Retrieved from EBSCOhost.

Breselor, S. (2010, September 23). Gordon Gekko's Cell Phone. *Salon.com*. Retrieved from http://img.slate.com/blogs/blogs/browbeat/archive/2010/09/23/gordon-gekko-s-cell-phone.aspx.

Brncik, M. (2011). A Case For Non-Enforcement of Anti-Recording Laws Against Citizen Recorders. *Case Western Reserve Law Review, 62*(2), 485–522.

Broom, S. (2011, September 20). 2 Prince George's County Police Officers indicted for beating after Maryland-Duke Game in College Park. *WUSA9.com*. Retrieved from http://www.wusa9.com/news/local/story.aspx?storyid=167773.

Browne, P. J. (2008, September 25). Statement by Deputy Commissioner Paul J. Browne, New York City Police Department. Retrieved from http://www.nytimes.com/2008/09/25/ nyregion/25taserletter.html.

Bulwa, D. (2010a, April 23). BART fires officer who ordered Grant arrested. *SFGate.com*. Retrieved from http://articles.sfgate.com/2010-04-23/bay-area/20862054_1_bart-force-officer-marysol-domenici-fruitvale-station.

Bulwa, D. (2010b, June 12). Grant's friend breaks down as his video plays. *SFGate.com*. Retrieved from http://www.sfgate.com/cgi-bin/article.cgi?f=/c/a/2010/06/12/ MN601DTUFC.DTL&ao=all.

Bulwa, D. (2010c, August 26). BART's price tag for Grant shooting: $6 million. *SFGate.com*. Retrieved from http://articles.sfgate.com/2010-08-26/bay-area/ 22235343_1_bart-shooting-marysol-domenici-oscar-grant.

Bulwa, D. (2010d, December 18). BART officer fired in Grant probe gets job back. *SFGate.com*. Retrieved from http://articles.sfgate.com/2010-12-18/bay-area/25207642_1_bart-officer-oscar-grant-marysol-domenici.

Bulwa, D., Fulbright, L., and Lee, H. K. (2009, January 14). BART officer arrested on murder warrant in NY Day shooting. *SFGate.com*. Retrieved from http://www.sfgate.com/cgi-bin/article.cgi?f=/c/a/2009/01/14/MNM615A08A.DTL.

Burger, C. (1955, October). Cover Your Town for TV. *Popular Photography*.

Burgess, J. E., & Green, J. B. (2008). Agency and controversy in the YouTube community. In *IR 9.0: Rethinking Communities, Rethinking Place* - Association of Internet Researchers (AoIR) conference. Presented at the IR 9.0: Rethinking Communities, Rethinking Place, Copenhagen, Denmark: unpublished. Retrieved from http://eprints.qut.edu.au/15383/.

Burgess, J., & Green, J. (2009). *YouTube: Online Video and Participatory Culture*. Cambridge; Malden, MA: Polity.
Burke, K, and White, M. (2007, May 23). 'Heartless' bust: 33 nabbed on way to wake. *New York Daily News*. Retrieved from http://www.nydailynews.com/news/crime/heartless-bust-33-nabbed-wake-article-1.253857.
Burke, K., and Connor, T. (2008, September 30). Family of Brooklyn man who died after being Tasered by cop to sue. The *New York Daily News*. Retrieved from http://articles.nydailynews.com/2008-09-30/local/29435302_1_iman-morales-cop-shot-mistake-in-taser-tragedy.
Burns, K. (2009 January 17). The E-List: St. Sen. Gladden and Prince George's spokesman Keary. *The Maryland News Company*. Retrieved from http://www.marylandpolitics.us/ archives/1047.
Byrnes, T. (1886). *Professional Criminals of America*. New York: Cassel.
California v. Mehserle, No. 547353 (2009a, June 4). Reporter's transcript of proceedings: Preliminary hearing. (6) 915. Retrieved from http://www.indybay.org/uploads/2009/07/ 17/mehserlepreliminaryhearingvol6part3_060409.pdf.
California v. Mehserle, No. 161210 (2009b, October 16). Order of the court on defendant's motion to change venue. Superior Court of California, County of Alameda.
California v. Mehserle, No. 161210 (2010, February 19). Trial Brief. Superior Court of California, County of Alameda.
Calvert, S., and Jones, B. (2010, April 13). Prince George's Co. police chief voices outrage over video of College Park beating. *The Baltimore Sun*. Retrieved from http://articles.baltimoresun.com/2010-04-13/news/bs-md-pg-officer-suspended-um-20100413_1_police-officers-prince-george-s-county-police-beating.
Cane, M. (2008a, February 12). YouTube justice: Officer Rivieri suspended. Retrieved from http://mikecane2008.wordpress.com/2008/02/12/youtube-justice-officer-rivieri-suspended/.
Cane, M. (2008b, February 15). Officer Rivieri fails again. Retrieved from http://mikecane2008.wordpress.com/2008/02/15/officer-rivieri-fails-again/.
Canon. (2011). Canon Camera Story 1976-1986. In *Canon Camera History Museum*. Retrieved from http://www.canon.com/camera-museum/history/canon_story/1976_1986/ 1976_1986.html.
Carlebach, M. L. (1992). *The Origins of Photojournalism in America*. Washington DC: Smithsonian Institution Press.

Carter, M. (2011, October 5). Man sues, says SPD conceals dash-cam videos. *The Seattle Times*. Retrieved from http://seattletimes.nwsource.com/html/localnews/ 2016416424_dashcam06m.html.

Carter, M., & Miletich, S. (2011, December 15). Justice Department report blasts Seattle police. *The Seattle Times*. http://seattletimes.nwsource.com/html/localnews/ 2017026414_doj16m.html.

Carter, R. L. (2007). Advancement of digital photography and related technologies timetable. In M. R. Peres (Ed.), *Focal Encyclopedia of Photography: Digital Imaging, Theory and Applications, History, and Science*. Burlington, MA: Focal Press.

Carter, R. L. (2011). 1988. *DigiCamHistory.Com*. Retrieved from http://www.digicamhistory.com/1988.html.

Castaneda, R. (2005, May 13). Reporter alleges excessive force: journalists are stopped while following Pr. George's official. *The Washington Post*. Friday final edition, Metro section, B05. Retrieved from LexisNexis Academic.

Castaneda, R. (2007, April 17). TV reporter sues over alleged police force, intimidation. *The Washington Post*. Tuesday suburban edition, Metro section, B06. Retrieved from LexisNexis Academic.

Castaneda, R. (2009, January 17). Police used excessive force, civil jury says; TV reporter awarded $5,000 in damages for '05 traffic stop. *The Washington Post*. Saturday suburban edition, Metro section, B06. Retrieved from LexisNexis Academic.

Castaneda, R. (2010a, April 12). Video evidence has spurred internal investigations in Pr. George's. *The Crime Scene* (blog). The Washington Post. Retrieved from http://voices.washingtonpost.com/crime-scene/charges-of-pr-georges-police-b.html.

Castaneda, R. (2010b, April 20). Video evidence has spurred internal investigations in Pr. George's. The Crime Scene (blog). *The Washington Post*. Retrieved from http://voices.washingtonpost.com/crime-scene/ruben-castaneda/charges-of-pr-georges-police-b.html.

Castaneda, R. (2010c, July 19). Officers' e-mails probed in U-Md. case. *The Washington Post*. Suburban Edition. Metro, B01. Retrieved from LexisNexis Academic News.

Castaneda, R. (2011, May 17). Johnson, ex-county executive in Prince George's, pleads guilty to taking bribes. *The Washington Post*. Retrieved from http://www.washingtonpost.com/ local/johnson-ex-county-executive-in-prince-georges-pleads-guilty-to-taking-bribes/2011/05/17/AF0pP75G_story_1.html.

Castells, M. (2007). Communication, power and counter-power in the network society. *International Journal of Communication, 1*, 238-266.

Cave, S. (n.d.). Grind. *Skateboarding Dictionary*. Retrieved from http://skateboard.about.com/od/skateboardingdictionary/g/GlosGrind.htm.

Center for Constitutional Rights. (2011, January 25). Amicus Brief in *Glik v. Cunniffe, et al.* Retrieved from http://ccrjustice.org/ourcases/current-cases/glik.

Chadwick, E. (1868, January). On the consolidation of the police force, and the prevention of crime. *Fraser's Magazine (67)*, 16.

Channel 10, Inc. v. Gunnarson, 337 F. Supp. 634 (D. Minn. 1972).

Chapman et al. v. City of New York et al., 10-cv-03067 (2011 January 10). Document 8: First amended complaint. Retrieved from PACER.

Chapman et al. v. City of New York et al., 10-cv-03067 (2011 January 10b). Document 8-A: First amended complaint exhibit A. Retrieved from PACER.

Chapman et al. v. City of New York et al., 10-cv-03067 (2011 April 11). Document 10: Answer to amended complaint on behalf of City of New York. Retrieved from PACER.

Chapman et al. v. City of New York et al., 10-cv-03067 (2011 May 10). Document 15: Answer to first amended complaint on behalf of Marshal Winston and Eugenia Williams. Retrieved from PACER.

Chapman et al. v. City of New York et al., 10-cv-03067 (2011 June 1). Document 16: Notice of settlement. Retrieved from PACER.

Chapman et al. v. City of New York et al., 10-cv-03067 (2011 June 6). Document 17: Stipulation of settlement and order of dismissal. Retrieved from PACER.

Chapman v. City of New York et al: Justia dockets & filings. (2011, May 17). Retrieved from http://dockets.justia.com/docket/new-york/nyedce/1:2011cv02370/317918/.

Chapman, J. (charmcityinc) (2008, February 12). Baltimore cop V.S. BMXers. Retrieved from http://www.youtube.com/watch?v=jsPXUu7WRTA.

Chavarria et al. v. City of New York et al., 1:08-cv-03034-ENV-RLM (2008, July 23). Document 1: Complaint against City of New York, Raymond Kelly, Scott Henderson, John Does, filed by Lanira Chavarria, Otima Martinez, Sandy Taveras. US District Court, Eastern District of New York. Retrieved from PACER.

Chavarria et al. v. City of New York et al., 1:08-cv-03034-ENV-RLM (2009, June 24). Document 17: Settlement agreement. US District Court, Eastern District of New York. Retrieved from PACER.

Cherry, R. (2010, October 29). Bull & Oyster Roast - P/O Sal Rivieri. Retrieved from http://www.fop3.org/index.php?option=com_content&view=article&id=178:bull-a-oyster-roast-po-sal-rivieri.

Cheshire, C., & Antin, J. (2008). The social psychological effects of feedback on the production of Internet information pools. *Journal of Computer-Mediated Communication, 13*(3), 705 - 727.

ChooseRonPaul. (2008, September 25). Cops in nude taser slay - Brooklyn death plunge. Retrieved from http://www.youtube.com/watch?v=Sg5RVq4m8xg.

Christopher, W., & Independent Commission on the Los Angeles Police Department (1991). *Report of the Independent Commission on the Los Angeles Police Department*. Los Angeles, CA: The Commission.

Chung, J. (2008, January 29). Brooklyn DA's office throws out mourners' arrests. *Gothamist*. Retrieved from http://gothamist.com/2008/01/29/brooklyn_das_of_1.php.

Chung, J. (2008, September 25). After cop Tasers him, naked man falls to death. *Gothamist*. Retrieved from http://gothamist.com/2008/09/25/after_cop_tasers_him_naked_man_fall.php.

Chung, J. (2008, October 3). ESU cop worried he "disgraced his family". *Gothamist*. Retrieved from http://gothamist.com/2008/10/03/esu_cop_worried_he_disgraced_his_fa.php.

Citizen Complaint Oversight Panel. (2010). *2010 Annual Report*. Prince George's County, MD. Retrieved from http://www.co.pg.md.us/Government/BoardsCommissions/ pdfs/2010-annual-report.pdf.

Citizens Police Review Board. (2007a). Citizens Police Review Board: Agenda/Notice. Fort Lauderdale, FL. Retrieved from http://ci.ftlaud.fl.us/documents/crb/crb_agenda.htm.

Citizens Police Review Board. (2007b, December 10). Agenda: Citizens Police Review Board meeting. Fort Lauderdale, FL. Retrieved from http://ci.ftlaud.fl.us/documents/crb/2007/ 121007cprb_agenda.pdf.

Citizens Police Review Board. (2007c, December 10). Minutes: Citizens Police Review Board meeting. Fort Lauderdale, FL. Retrieved from http://ci.ftlaud.fl.us/documents/crb/2007/ 121007cprb_minutes.pdf.

City of Los Angeles v. Lyons (1983). Supreme Court of the United States. No. 81-1064. 461 U.S. 95.

City of Newark. (2011, August 3). Resolution: Settlement of Civil Litigation Docket No.: 09-5396 (FSH-PS). File 11-0807. Newark, NJ. Retrieved from http://newark.legistar.com/ LegislationDetail.aspx?ID=927786&GUID=737CED6C-0BD0-41DC-AEEB-F01BEBD5B46F.

Classically Liberal. (2008, February 15). Update on officer dude. [blog]. Retrieved from http://freestudents.blogspot.com/2008/02/update-on-officer-dude.html.

Clock, M. (2009, November 14). MTS begins expanding camera surveillance at trolley stations. *Sign On San Diego*. Retrieved from http://www.signonsandiego.com/news/2009/nov/ 14/mts-begins-expanding-camera-surveillance-at/.

CNN. (2010, May 25). Cops charge man for YouTube post. *American Morning*. Retrieved from http://www.youtube.com/watch?v=2YSx2LjXJZo.

Cobus, P. (2008, February 20). Lessons from 'cop vs. skater'. The *Baltimore Sun*, A9.

Collins, D. (1990). *The Story of Kodak*. New York: H.N. Abrams.

Columbia Human Rights Law Review. (1973). *Surveillance, dataveillance, and personal freedoms: Use and abuse of information technology*. A symposium edited by the staff of Columbia Human Rights Law Review. Fair Lawn, NJ; R. E. Burdick.

Compton, M. (2000, May 22). Camera on a chip: Photobit CEO Sabrina Kemeny. *Salon.com*. Retrieved from http://archive.salon.com/tech/view/2000/05/22/kemeny/index.html.

Craig, D. (2011). Vintage camcorders. *Rewind Museum*. Vision International. Retrieved from http://www.rewindmuseum.com/vintagecamcorder.htm.

Cramer, M. (2012, January 10).Police reverse stance on taping of officers' actions. *The Boston Globe*. Retrieved from http://www.bostonglobe.com/metro/2012/01/10/police-reverse-stance-taping-officers-actions/va6glfwq9L1mUElv6a33HK/story.html.

Crapsey, E. (1872). The Nether Side of New York: The Vice, Crime and Poverty of the Great Metropolis. New York, NY: Sheldon.

Creswell, J. W. (2007). *Qualitative Inquiry & Research Design: Choosing Among Five Approaches*. (2nd ed.). Thousand Oaks: Sage Publications.

Cruz, W. (2009, April 19). City must cough up 257G for false arrest of Bushwick 32. *New York Daily News*. Retrieved from http://www.nydailynews.com/new-york/city-cough-257g-false-arrest-bushwick-32-article-1.361632.

Cruz, W. (2009, June 12). "Bushwick 32" get vindication: City settles with Brooklyn teens arrested on way to friends' wake. *New York Daily News*. Retrieved from http://www.nydailynews.com/new-york/brooklyn/bushwick-32-vindication-city-settles-brooklyn-teens-arrested-friends-wake-article-1.378610.

Cummings, M. (2010, July 13). Brooklyn women knew their rights — and you should too! *Blog of Rights: Official Blog of the American Civil Liberties Union*. Retrieved from http://www.aclu.org/blog/immigrants-rights-racial-justice/brooklyn-women-knew-their-rights-and-you-should-too.

Cunningham, L. A. (2002). Outsmarting the Smart Money : Understand how Markets Really Work and Win the Wealth Game. New York: McGraw-Hill.

Curlin, C. (2010, April 26). Police adjust to a world caught on tape. *Washington Times, The* (DC), 1. Washington, D.C.

Curran, J. J. (2000, August 11). Wiretap and electronic surveillance. Retrieved from http://www.oag.state.md.us/Opinions/2000/85oag225.pdf.

Curry, T. (1985). Frederick Wiseman: Sociological Filmmaker? *Contemporary Sociology*, *14*(1), 35-39.

CW11. (2008, September 24). Nude Man Tasered Killed by Police. Retrieved from YouTube channel policecrimedotcom http://www.youtube.com/v/OiXy9WqeLRQ and http://www.veoh.com/watch/v16063042xapsn28c?.

de Gennaro, S. (n.d.). LHC: The First Band on the Web. *CERN Musiclub*. Retrieved from http://musiclub.web.cern.ch/MusiClub/bands/cernettes/firstband.html.

de Graaf, G., & Huberts, L. W. J. C. (2008). Portraying the nature of corruption using an explorative case study design. *Public Administration Review, 68*(4), 640-653.

Del Signore, J. (2008, October 1). Tasered man's family hires lawyer Ron Kuby. *Gothamist*. Retrieved from http://gothamist.com/2008/10/01/tasered_mans_family_will_sue.php.

DeMetrick, A. (2011, February 28). Firing Of Officer Seen In YouTube Video Stands. WJZ. Retrieved from http://baltimore.cbslocal.com/2011/02/28/firing-of-officer-seen-in-youtube-video-stands/.

Denney, R. (1955). The plainest plain-clothesman. *New Republic, 132*(5), 14-16.

Detroit News (1917). *Detroit News Newsreel*. Detroit: Prelinger Archives. Retrieved from http://www.archive.org/download/DetroitN1917/DetroitN1917.mpeg.

Diaz, J. (2011, August 30). Message to the community from Seattle Police Chief John Diaz. Retrieved from http://www.kirotv.com/news/news/message-to-the-community-from-seattle-police-chief/nDkqZ/.

Diehl, R. N. (2005, January 6). 1965 Sony VCK-2000 1 Vidicon Camera Kit for the CV-2000 Series. *Labguy's World*. Retrieved from http://www.labguysworld.com/Sony_VCK-2000.htm.

Directing a Stranger. (1899). [photographic print, 7 1/2 x 5 1/4 in.]. Picture Collection (image ID 806226, barcode 33333159171558, Struc ID 577561). New York Public Library, Mid-Manhattan Library, 455 Fifth Avenue, New York, NY. Retrieved from http://digital.nypl.org/mmpco/imageidresults.cfm?trg=1&imageid=80622.

DJacobs2009. (2009, February 20). Police Misconduct. *YouTube*. Retrieved from http://www.youtube.com/watch?v=2uEUxbTHgt0.

Dorcey, T. (1995, March). CU-SeeMe Desktop VideoConferencing Software. *Connexions, 9*(3). Retrieved from http://ipsix.org/source/dorcey.html.

downsouthvids. (2008, July 27). Critical Mass Bicyclist Assaulted by NYPD. *YouTube*. Retrieved from http://www.youtube.com/watch?v=oUkiyBVytRQ.

Doyle, A. (2003). Arresting Images: Crime and Policing in Front of the Television Camera. Toronto: University of Toronto Press.

Doyle, J., Schram, J., and Lenkowitz, E. (2008, September 25). Cops in nude taser slay: Brooklyn death plunge after 2nd-story jolt. The *New York Post*. Regional news. Retrieved from http://www.nypost.com/seven/09252008/news/regionalnews/cops_in_nude_taser_slay_130670.htm.

Dubber, M. D. (2006). Victims in the War on Crime: The Use and Abuse of Victims' Rights. New York: New York University Press.

Duggan, M. (2013, October 28). Photo and Video Sharing Grow Online. Pew Research Center. Retrieved from http://pewinternet.org/Reports/2013/Photos-and-videos.aspx.

Durkin, K. (2007, June 14). Victimized youth organize against police brutality. *International Action Center - Charlotte*. Retrieved from http://charlotteaction.blogspot.com/ 2007/06/victimized-youth-organize-against.html.

Dwoskin, E. (2009a, April 20). City agrees to pay 250K to wrongly-arrested Bushwick kids. *Village Voice*. Retrieved from http://blogs.villagevoice.com/runninscared/ 2009/04/ city_agrees_to.php?.

Dwoskin, E. (2009b, July 28). New York stop-and-frisk 101: Arrest training? *Village Voice*. Retrieved from http://www.villagevoice.com/content/printVersion/1279151/.

Eastman Kodak Company. (1999, March). *History Of Kodak cameras: Customer service pamphlet*. Kodak Publication No. AA-13. Retrieved from http://www.kodak.com/ global/en/consumer/products/techInfo/aa13/aa13.pdf.

Eastman Kodak Company. (n.d.). *History of Kodak: Milestones - Chronology*. Retrieved from http://www.kodak.com/global/en/corp/historyOfKodak/chronology.jhtml.

Eddlem, T. R. (2010). The Power of the Internet. (Cover story). *New American (08856540)*, *26*(18), 10-14.

Edgar, J. (2011, July 19). Newark Police - "I can do whatever I want". *New Jersey Libertarian Party*. Retrieved from http://njlp.org/news/partynews/policeaccountability/1199-newark-police-arrest-cameraman.

EducatedBlackGirl [Williams, M.]. (2010, July 11). Two [Br]ooklyn women claim they were wrongfully arrested by the NYPD after following the advice of a flyer entitled, "What should you do if stopped by the police." [Online forum message]. Retrieved from http://www.nydailynews.com/forums/thread.jspa?threadID=102917.

Edwards, P., & Edwards, S. (1999). The Best Home Businesses For the 21st Century: The Inside Information You Need to Know to Select a Home-Based Business That's Right for You. New York: Jeremy P. Tarcher/Putnam.

Efrati, A. (2009, January 29). Legal system struggles with how to react when police officers lie. *Wall Street Journal*. A12.

Efrati, A. (2012, April 13). Google stock move puts it in unusual company. Digits (blog). *The Wall Street Journal*. Retrieved from http://blogs.wsj.com/digits/ 2012/04/13/google-stock-move-puts-it-in-unusual-company/.

Eggerton, J. (2007 April 16). WJLA reporter sues over traffic stop. *Broadcasting & Cable*. Retrieved from http://www.broadcastingcable.com/article/108544-WJLA_Reporter_Sues_Over_Traffic_Stop.php.

Ehrlich, K. (2010 August 28). WBAL exclusive: Officer Rivieri speaks out on firing. [Audio recording]. WBAL, Baltimore MD. Retrieved from http://www.wbal.com/absolutenm/ articlefiles/57608-OFFICER%20RIVIERI%20CALLING%20KENDEL%20EHRLICH%20SHOW%20SATURDAY%20AUG%2028%202010.mp3.

Eligon, J. & Moynihan, C. (2008, December 16). Police officer seen on tape shoving a bicyclist is indicted. *The New York Times*. Metropolitan Desk, A0. 33. Retrieved from LexisNexis Academic News.

Eligon, J. (2010, April 30). Former officer is found guilty of lying about confrontation with bicyclist. *The New York Times*. Late Edition – Final. Metropolitan Desk, A0. 19. Retrieved from LexisNexis Academic News.

Emery, C. (2008, February 16). Strife with police is old; Clash with authority familiar to skaters. *The Baltimore Sun*, B1.

Epstein, V. (2008, October 27). Officer suspended over cameraman arrest in Newark. State and regional. *The Associated Press State & Local Wire*. Retrieved from LexisNexis Academic.

EverReady Battery Company, Inc. (2001, November 6). *Eveready Carbon Zinc (Zn/MnO²) Application Manual*. Retrieved from http://data.energizer.com/PDFs/ carbonzinc_appman.pdf.

Exzellent. (2008, February 17). Skateboard-hating Baltimore cop at it again. Retrieved from http://www.youtube.com/watch?v=FedlpshJb4c.

Facebook. (2011, December 27). I support Anthony Graber and his right to freedom of expression. Retrieved from http://www.facebook.com/pages/I-support-Anthony-Graber-and-his-right-to-freedom-of-expression/120203541353801?sk=wall.

Faheem, K. and Hauser, C. (2008, September 26). Taser use in man's death broke rules, police say. *The New York Times*. Retrieved from LexisNexis Academic.

Fairchild Imaging. (n.d.). CCD, CMOS, OEM CCD Cameras - About Us - History. *Fairchild Imaging*. Retrieved from http://www.fairchildimaging.com/about/history.htm.

Farber, D. (2003). *Chicago '68*. Chicago, IL: University of Chicago Press.

Farrell, L. (2010, April 30). Social media changes the face of political activism in New Jersey and beyond. Examiner.com. Retrieved from: http://www.examiner.com/social-media-in-newark/social-media-changes-the-face-of-political-activism-new-jersey-and-beyond.

Fenton, J. (2008, December 12). Judge lets skateboarder lawsuit go ahead. A4. *The Baltimore Sun*.

Fenton, J. (2009, January 7). In policy change, city police will not identify officers who kill, injure. A1. *The Baltimore Sun.*

Fenton, J. (2012, January 11). DOJ urges judge to side with plaintiff in Baltimore police taping case. *The Baltimore Sun.* Retrieved from http://www.baltimoresun.com/news/breaking/ bs-md-ci-aclu-doj-videotaping-20120111,0,7691935.story.

Ferdinand, T. N. (1980). Criminality, the courts, and the constabulary in Boston: 1702-1967. *Journal of Research in Crime and Delinquency, 17*(2), 190-208.

Fielding, R. (2006). *The American Newsreel: A Complete History, 1911-1967.* Second edition. Jefferson, NC: McFarland.

Fildes, S. L. (1873, November 29). The Bashful Model. *Harper's Weekly.* 1064-1065. Retrieved from: http://hdl.loc.gov/loc.pnp/cph.3c07510.

Finisterre, K. (2011, April). Owning a Cop Car. Digitalmunition.com. Retrieved from http://www.digitalmunition.com/OwningCopCar.pdf.

Flickr: Photography is not a crime. (n.d.). . Retrieved April 24, 2010, from http://www.flickr.com/groups/photography_is_not_a_crime/.

Fogelson, R. M. (1977). *Big-city Police.* Cambridge, Mass. Harvard University Press.

Fong, K. (2012, January 5). Notice of finding. Boston Police Department. Retrieved from http://www.rcfp.org/sites/default/files/docs/20120112_173329_boston_police_notice_to_simon_glik.pdf.

Fordyce v. City of Seattle, 55 F.3d 436 (9th Cir. 1995).

Fort Lauderdale Police Department. (n.d.). City of Fort Lauderdale Police Department Internal Affairs. Fort Lauderdale, FL. Retrieved from http://www.flpd.org/index.aspx?page=51.

Fossum, E. R. (1995). CMOS image sensors: Electronic camera-on-a-chip. *IEEE International Electron Devices Meeting Digest of Technical Papers.* 17.

Fossum, E. R. (1997, October). CMOS image sensors: Electronic camera-on-a-chip. *IEEE Transactions on electron devices, 44*(10), 1689.

Fossum, E. R. (2011). Biography of Eric R. Fossum. Retrieved from http://ericfossum.com/ Resume/Extended Biography.htm.

FOX 25. (2010, December 22). Training video could be used against Boston Police in lawsuit. MyFoxBoston.com. Retrieved from http://www.myfoxboston.com/dpp/news/local/ training-video-could-be-used-against-boston-police-in-lawsuit-20101222.

Fox News. (2009 January 6). Man's fatal shooting by San Francisco subway police prompts probe, plans for $25M lawsuit". Retrieved from http://www.foxnews.com/story/ 0,2933,476644,00.html.

Fox, S. (2011). *Health Topics*. Pew Internet & American Life Project (p. 33). Washington, D.C.: Pew Research Center. Retrieved from http://www.pewinternet.org/~/media//Files/ Reports/2011/PIP_HealthTopics.pdf.

Frank, D. E. (2008, February 4). Wiretap case against lawyer under advisement. *Massachusetts Lawyers Weekly*. Retrieved from http://masslawyersweekly.com/2008/02/04/wiretap-case-against-lawyer-under-advisement/.

Frankfurter Allgemeine Zeitung GmbH. Informationsdienste. (1991). *German brief, 3*, 24.

Friebele, B. (2011) Statement. Art by Billy Friebele. Retrieved from http://www.billyfriebele.com/ArtistStatement.html.

FunkensteinJr. (2010). Cop pulls out gun on motorcyclist. *YouTube*. Retrieved from http://www.youtube.com/watch?v=RK5bMSyJCsg .

Galambos, L., & Abrahamson, E. J. (2002). *Anytime, Anywhere: Entrepreneurship and the Creation of a Wireless World*. Cambridge, UK; New York: Cambridge University Press.

Ganging up on cops. (2007, May 30). *New York Post*, p. 32. New York, NY.

Gannett Satellite Info. Network, Inc. v. Township of Pennsauken, 709 F. Supp. 530, 536-37 (D.N.J. 1989).

Gartner Dataquest. (2006). Gartner Says Nearly 50 Percent of Worldwide Mobile Phones Will Have a Camera in 2006 and 81 percent by 2010. Press release. Retrieved from http://www.gartner.com/it/page.jsp?id=498310.

Gecker, J. (2010, April 20). Hotels in Bangkok sending tourists packing. Associated Press. Retrieved from http://www.google.com/hostednews/ap/article/ ALeqM5hrELLmLT354u1ruuAvJu-EHC6CFgD9F6TUUO1.

Gendar, A. (2008, September 30). Police Commissioner Kelly admits 'mistake' in Taser tragedy. The *New York Daily News*. Retrieved from http://www.nydailynews.com/news/ny_crime/ 2008/09/29/2008-09-29_police_commissioner_kelly_admits_mistake.html.

Giles, B. (2011, September 20). Two P.G. cops indicted in beating of U.Md. student. Washington Examiner. Retrieved from http://washingtonexaminer.com/local/2011/09/ two-pg-cops-indicted-beating-umd-student?quicktabs_1=0.

Gillespie, T. (2010). The politics of 'platforms'. *New Media & Society, 12*(3), 347-364.

Gilligan, J. (1958, January). Television documentaries use amateur footage. *Popular Photography*. 120.

GioSifaTaufa. (2009, January 4). Oakland, CA Fruitvale BART station shooting. Full video and interview with Karina Vargas. *YouTube*. Retrieved from http://www.youtube.com/ watch?v=s_K0PISpxx4.

Given, L. M. (Ed.). (2008). *The SAGE Encyclopedia of Qualitative Research Methods*. Los Angeles, CA: SAGE Publications.

Glass Bead Collective. (2007, March 30). *Some Assembly Required*. Retrieved from http://www.glassbeadcollective.org/cm0307.htm.

Glik v. Cunniffe et al. (2010). Civil Action No. 10-10150. United States District Court, District of Massachusetts.

Glik v. Cunniffe et al., No. 1:10-cv-10150-WGY (2010a, February 1). Document 1: Complaint. Retrieved from PACER.

Glik v. Cunniffe et al., No. 1:10-cv-10150-WGY (2010b, February 5). Document 6: Amended complaint. Retrieved from PACER.

Glik v. Cunniffe et al., No. 1:10-cv-10150-WGY (2010c, June 23). Document 22: Notice of appeal. Retrieved from PACER.

Glik v. Cunniffe et al., No. 1:10-cv-10150-WGY (2010f, August 10). Document 30: Transcript of motion hearing. Retrieved from PACER.

Glik v. Cunniffe et al., No. 1:10-cv-10150-WGY (2010g, August 27). Document 34: Order to stay the District Court proceedings. Retrieved from PACER.

Glik v. Cunniffe et al., No. 1:10-cv-10150-WGY (2011f, August 29). Document 44: Judgment. Retrieved from PACER.

Glik v. Cunniffe et al., No. 1:10-cv-10150-WGY (2011g, November 7). Document 54: Defendant City of Boston's supplemental memorandum in support of its motion to maintain the "confidential—subject to protective order" designation of certain documents produced through discovery. Retrieved from PACER.

Glik v. Cunniffe et al., No. 10-1764 (2010d, June 29). Case opening notice. Appealed from 1:10-cv-10150-WGY. Retrieved from PACER.

Glik v. Cunniffe et al., No. 10-1764 (2010e, August 4). Order to stay the District Court proceedings. Appealed from 1:10-cv-10150-WGY. Retrieved from PACER.

Glik v. Cunniffe et al., No. 10-1764 (2010h, December 13). Brief of appellants John Cunniffe, Peter Savalis, and Jerome Hall-Brewster. Appealed from 1:10-cv-10150-WGY. Retrieved from PACER.

Glik v. Cunniffe et al., No. 10-1764 (2011a, January 18). Brief of plaintiff-appellee Simon Glik. Appealed from 1:10-cv-10150-WGY. Retrieved from PACER.

Glik v. Cunniffe et al., No. 10-1764 (2011b, January 25). Motion for leave to file *Amicus curiae* brief. Appealed from 1:10-cv-10150-WGY. Retrieved from PACER.

Glik v. Cunniffe et al., No. 10-1764 (2011c, February 9). *Amicus curiae* brief filed by movants Berkeley Copwatch, Communities United Against Police Brutality, Justice Committee, Milwaukee Police Accountability Coalition, Nodutdol for Korean Community Development and Portland Copwatch in support of Appellee. Appealed from 1:10-cv-10150-WGY. Retrieved from PACER.

Glik v. Cunniffe et al., No. 10-1764 (2011d, April 19). Corporate disclosure statement filed by Amici Curiae Berkeley Copwatch, Citizen Media Law Project, Communities United Against Police Brutality, Justice Committee, Milwaukee Police Accountability Coalition, Nodutdol for Korean Community Development and Portland Copwatch. Appealed from 1:10-cv-10150-WGY. Retrieved from PACER.

Glik v. Cunniffe et al., No. 10-1764 (2011e, August 26). Order. Appealed from 1:10-cv-10150-WGY. Retrieved from http://www.ca1.uscourts.gov/pdf.opinions/10-1764P-01A.pdf.

Glik, S. (2007, October 1). 1001071733.3g2. (video clip). Retrieved from http://www.aclum.org/ news/video/1001071733.3g2.

Godfrey, D. G. (2001). *Philo T. Farnsworth: The Father of Television.* Salt Lake City, UT: University of Utah Press.

Goldstein, A. (2007, January 2). The private arm of the law. *Washington Post.* Retrieved from: http://www.washingtonpost.com/wp-dyn/content/article/2007/01/01/ AR2007010100665_pf.html.

Goldstein, A. (2008). Privacy from photography: Is there a right not to be photographed under New York state law? *Cardozo Arts & Entertainment Law Journal*, 26, 233-267.

Gonsalvez, A. (2010, April 7). Android Phones Steal Market Share. *InformationWeek.* Retrieved from http://www.informationweek.com/news/224201881#.

Gonzales, R. (2010, July 9). Justice Department to investigate BART shooting. *National Public Radio.* Retrieved from http://www.npr.org/templates/story/story.php?storyId=128409603.

Good Morning America. (2008, February 13). Officer Suspended After Skateboarder Rant. Retrieved from http://abcnews.go.com/GMA/story?id=4282823&page=1.

Google. (2011a). Transparency Report: United States. Retrieved from http://www.google.com/transparencyreport/governmentrequests/US/.

Google. (2011b, October). Transparency report: Government requests. Retrieved from http://www.google.com/transparencyreport/governmentrequests/.

Gordon, C., and Stabley, M. (2011, September 21). Two cops indicted in alleged UMD beating. Retrieved from http://www.nbcwashington.com/news/breaking/Two-Cops-Indicted-in-UMD-Alleged-UMD-Beating-130225438.html.

Gould, J., and Gendar, A. (2008, September 27). ESU officers to get new training in dealing with mentally disturbed. The *New York Daily News*. Retrieved from http://www.nydailynews.com/news/2008/09/26/2008-09-26_esu_officers_to_get_new_training_in_deal.html.

Graber, A. J. (nikotyc) (2008). nikotyc's channel. *YouTube*. Retrieved from http://www.youtube.com/user/nikotyc/feed.

Graber, A. J. (nikotyc) (2010a March 10). Motorcycle traffic violation - Cop pulls out gun. *YouTube*. Retrieved from http://www.youtube.com/watch?v=BHjjF55M8JQ.

Graber, A. J. (nikotyc) (2010b March 12). Motorcycle traffic violation - Cop pulls out gun (extended no sound). *YouTube*. Retrieved from http://www.youtube.com/watch? v=G7PC9cZEWCQ&NR=1.

Grace, M. (2010a, April 23). Bike-hitting cop went rogue: Sgt. *Daily News*. Sports final edition. News section. 53. New York, NY. Retrieved from LexisNexis Academic News.

Grace, M. (2010b, April 27). Biker's lawyer brands ex-cop a liar who's 'covering up assault'. *Daily News*. Sports final edition. News section. 15. New York, NY. Retrieved from LexisNexis Academic News.

Grace, M. (2010c, April 30). Bike cop's split decision: Officer found guilty of lies, not of cyclist's assault. *Daily News*. Sports final edition. News section. 8. New York, NY. Retrieved from LexisNexis Academic News.

Grant, O. (2009, January 1). Photograph of BART Police Officer Johannes Mehserle. Retrieved from http://www.indybay.org/uploads/2010/06/30/oscar-grant-mehserle-trial-oscars-cell-phone-pic-of-mehserle-provided-by-la-superior-ct.jpg.

Green, Jr. et al. v. The City of New York et al., 1:08-cv-02240-ENV-RLM (2008, June 4). Document 1: Complaint against "Richard Roes One through Twelve", The City of New York, Raymond Kelly, Scott Henderson, filed by Daryl Green, Jr., Stefhon Jackson, Ivan Morgan. US District Court, Eastern District of New York. Retrieved from PACER.

Green, Jr. et al. v. The City of New York et al., 1:08-cv-02240-ENV-RLM (2009, July 6). Document 28: Settlement agreement. US District Court, Eastern District of New York. Retrieved from PACER.

Green, S. J. (2012, June 27). SPD lawsuit over racial slur settled for $150K. *seattletimes.com*. Retrieved from http://seattletimes.com/ html/localnews/2018541588_monetti28m.html.

Greenough, S. (2007). The Art of the American Snapshot, 1888-1978: From the Collection of Robert E. Jackson (Hardcover ed.). Washington DC: National Gallery of Art.

GSMArena Team. (2010, May 19). Nokia N8 in focus: An interview with Damian Dinning. *GSM Arena*. Retrieved from http://www.gsmarena.com/nokia_n8_camera-review-476.php.

Guilfoil, J. M. (2010, February 1). ACLU files suit over cellphone video of police. Metro Desk. *Boston.com*. Retrieved from http://www.boston.com/news/local/breaking_news/2010/ 02/_wwwbostoncomne.html.

Gustavson, T. (2009). Camera: A History of Photography from Daguerreotype to Digital. New York: Sterling Innovation.

Gye, L. (2007). Picture this: the impact of mobile camera phones on personal photographic practices. *Continuum: Journal of Media & Cultural Studies, 21*(2), 279-288.

Gye, L. (2008). Picture this: the impact of mobile camera phones on personal photographic practices. In G. Goggin (Ed.), *Mobile Phone Cultures* (pp. 135-142). London: Routledge.

Habtemariam, Z. (2010, June 3). Missing riot video deemed an oversight: Univ. Police cleared of wrongdoing. *The Diamondback*. Retrieved from http://www.diamondbackonline.com/ news/missing-riot-video-deemed-an-oversight-1.1488404.

Hain, A. (2011). JVC GR-C1. In *Total Rewind: The Virtual Museum of Vintage VCRs*. Retrieved from http://www.totalrewind.org/cameras/C_GRC1.htm.

Hakim, S., & Blackstone, E. A. (1997). *Securing Home and Business: A Guide to the Electronic Security Industry*. Boston: Butterworth-Heinemann.

Halbfinger, D. M. (2009, April 12). At New School Protest, Truth Depends on Camera Angle. *The New York Times*. (Late Edition (East Coast)). A24. Retrieved from ProQuest.

Hales, P. B. (1984). Silver Cities: The Photography of American Urbanization, 1839-1915. Philadelphia: Temple University Press.

Halsne, C. (2010a, May 6). Seattle cops stomp on detainee. *KIROTV.com*. Retrieved from http://www.kirotv.com/news/news/seattle-cops-stomp-on-detainee/nDRHc/.

Halsne, C. (2010b, May 6). Watch it: Officer threatens, kicks detainee. *KIROTV.com*. Retrieved from http://www.kirotv.com/videos/news/watch-it-officer-threatens-kicks-detainee/vCk6c/.

Halsne, C. (2012, February 21). Disciplined officer earns extra overtime. *KIROTV.com*. Retrieved from http://www.kirotv.com/news/news/disciplined-officer-earns-extra-overtime/nJQqG/.

Hannigan, W., & Johnston, K. (2004). *Picture Machine: the Rise of American Newspictures*. New York: Harry N. Abrams.

Harring, S. L. (1983). *Policing a Class Society: The Experience of American Cities, 1865-1915.* Crime, law, and deviance series. New Brunswick, NJ: Rutgers University Press.

Harris, D. (2009). What criminal law and procedure can learn from criminology: How accountability-based policing can reinforce-or replace-the fourth amendment exclusionary rule. *The Ohio State Journal of Criminal Law*. Fall 2009. 7 Ohio St. J. Crim. L. 149.

Harris, J. (2009). Always-on reality. *Backbone*, 9.

Harris, S. D. (2008, July 13). Mercury News interview: SanDisk CEO helped launch digital revolution. *Mercury News*. San Jose, CA. Retrieved from http://www.mercurynews.com/ ci_9868280.

Harrison, D., Doyle, J., Schram, J., Lenkowitz, E., and Celona, L. (2008, September 24). Cops in nude taser slay: Brooklyn death plunge after 2nd-story jolt. The *New York Post*. Regional news. Retrieved from http://www.youtube.com/watch?v=Sg5RVq4m8xg.

Harvard Law Review. (1974, May). The Rights of the Public and the Press to Gather Information. *Harvard Law Review, 87*(7) 1505-1533.

Harvard Law Review. (1998). Privacy, photography, and the press. *Harvard Law Review, 111*(4), 1086.

Harvard Law Review. (1999). Privacy, technology, and the California `anti-paparazzi' statute. *Harvard Law Review, 112*(6), 1367.

Harvard Law Review. (2013a). Civil Procedure - Class Actions - Southern District of New York Certifies Class Action Against City Police for Suspicionless Stops and Frisks of Blacks and Latinos. -- *Floyd v. City of*

New York, 82 Fed. R. Serv. 3d (West) 833 (S.D.N.Y. 2012). (2013). *Harvard Law Review, 126*(3), 826–833.

Harvard Law Review. (2013b). Seventh Circuit Preliminarily Enjoins Eavesdropping Law as Applied to Police Monitoring Program. *Harvard Law Review, 126*, 1162-1169.

Hayhtio, T., & Rinne, J. (2009). Little brothers and sisters are watching. *Information, Communication & Society, 12*(6), 840 - 859.

Haynes, S. L. (2009). Controlling Photography in (and of) Private Places. Retrieved from http://www.shaynes.com/temp/Controlling_Photography.pdf.

Hentoff, N. (2010, July 27). Hey, Ray Kelly, frisk this. *The Village Voice*, Columns, p. 1. New York, NY.

Herbert, B. (2007a, May 26). Arrested while grieving. *The New York Times*. p. A13. New York, NY. Retrieved from ProQuest Newsstand.

Herbert, B. (2007b, May 29). Small incidents are creating a big problem with the N.Y.P.D. *The New York Times*, p. A.19. New York, NY. Retrieved from ProQuest Newsstand.

Herbert, B. (2007c, June 26). Doubting the police. *The New York Times*. New York, NY.

Herbert, B. (2008, February 16). Cruel and gratuitous. *The New York Times*, p. A19. New York, NY. Retrieved from ProQuest Newsstand.

Herbert, B. (2009, April 18). No cause for arrest. *The New York Times*. New York, NY.

Hermann, P. (2009a, January 8). Secret police?. *The Baltimore Sun*. Retrieved from http://weblogs.baltimoresun.com/news/crime/ blog/2009/01/withholding_names_of_officers.html.

Hermann, P. (2009b, March 3). Baltimore Police on YouTube. *The Baltimore Sun*. Retrieved from http://weblogs.baltimoresun.com/ news/crime/blog/2009/03/ baltimore_police_on_youtube.html.

Hermann, P. (2009c, September 22). Harbor cop who yelled at skateboarder gets suit dismissed. *The Baltimore Sun*. Retrieved from http://weblogs.baltimoresun.com/news/crime/blog/ 2009/09/harbor_cop_who_yelled_at_skate.html.

Hermann, P. (2009d, September 23). Cop in Harbor skateboard incident says he was threatened. Baltimore crime beat. *The Baltimore Sun*. Retrieved from http://weblogs.baltimoresun. com/ news/crime/blog/2009/09/cop_in_harbor_skateboard_incid.html.

Hermann, P. (2009e, September 27). Skateboard clip's lesson fuzzy. Crime beat. *The Baltimore Sun.* Retrieved from http://articles. baltimoresun.com/2009-09-27/news/ 0909260139_1_file-cop-video.

Hermann, P. (2009f, September 28). Skateboarder versus cop. Baltimore crime beat. *The Baltimore Sun.* Retrieved from http://weblogs.baltimoresun.com/news/crime/blog/2009/09/ skateboarder_versus_cop.html.

Hermann, P. (2010a, July 20). City cop who berated skateboarder cleared. Baltimore crime beat. *The Baltimore Sun.* Retrieved from http://weblogs.baltimoresun.com/news/crime/blog/2010/07/ city_cop_who_berated_skateboar.html.

Hermann, P. (2010b, August 25). Baltimore cop who berated skateboarder fired. Baltimore crime beat. *The Baltimore Sun.* Retrieved from http://weblogs.baltimoresun.com/news/crime/ blog/2010/08/baltimore_cop_who_berated_skat.html.

Hermann, P. (2010c, August 25). Police union fights back for fired officer in Harbor incident. Baltimore crime beat. *The Baltimore Sun.* Retrieved from http://weblogs.baltimoresun.com/news/crime/ blog/2010/08/police_union_fights_back_for_f.html.

Hermann, P. (2010d, August 26). City police union calls firing of cop unfair. Baltimore crime beat. The *Baltimore Sun.* Retrieved from http://weblogs.baltimoresun.com/ news/crime/blog/2010/08/city_police_union_calls_firing.html.

Hermann, P. (2010e, August 27). Fired officer in skateboard incident gets support. Baltimore crime beat. *The Baltimore Sun.* Retrieved from http://weblogs.baltimoresun.com/news/crime/blog/2010/08/ fired_officer_in_skakeboard_in.html.

Hermann, P. (2010f, September 12). Public turns fickle about cop. Final edition. Local, 6A *The Baltimore Sun.*

Hermann, P. (2010g, September 27). Judge says man had right to tape traffic stop. *The Baltimore Sun.* Local, A.1.

Hermann, P. (2011, February 28). Judge upholds firing of officer in harbor skateboard incident. Baltimore crime beat. The *Baltimore Sun.* Retrieved from http://articles.baltimoresun.com/2011-02-28/news/bs-md-harbor-cop-firing-20110228_1_salvatore-rivieri-officer-rivieri-video-berating.

Hesseldahl, A. (1999, November 8). The Legacy of DSP1. *Electronic News.* Retrieved from http://findarticles.com/p/articles/ mi_m0EKF/is_45_45/ai_57398570/.

Hintz, E. S. (2009). Portable Power: Inventor Samuel Ruben and the birth of Duracell. *Technology & Culture, 50*(1), 24-57.

Hogarty, D. (2007, June 23). Paying your respects, getting no respect. *Gothamist*. Retrieved from http://gothamist.com/2007/06/23/ paying_your_res.php.

Holmes, P. S. (2010, December 14). Seattle City Attorney re: Seattle police stomping incident. Retrieved from http://www.kirotv.com/ news/news/full-text-of-seattle-city-attorneys-decision-on-st/nDNn2/.

Horan, D. P., & St. Bonaventure University. (2005). *A Legal Handbook for Freelance Photographers*. St. Bonaventure, NY: St. Bonaventure University.

Hornberger v. Am. Broad. Cos., 799 A.2d 566, 626-27 (N.J. 2002).

Howell, T. (2009, November 3). Hackettstown cameraman sues city of Newark. *New Jersey Herald*. Newark, NJ. Retrieved from http://www.njherald.com/story/news/hackettstown.

Howett, D. (2006). Television Innovations: 50 Technological Developments. Tiverton: Kelly.

Human Rights Watch. (1998, June). *Shielded from justice: Police brutality and accountability in the United States*. Retrieved from: http://www.hrw.org/legacy/reports/reports98/police/ uspo73.htm.

Humiston v. Universal Film Manufacturing Co., 189 App. Div. 467, 472, 178. N.Y. Supp. 752, 756 (lst Dep't 1919).

Hurlbut, R. (2009a, September 7). San Diego Trolley Police Prohibit Photography of Brutality. *YouTube.com*: Theworldisraw. Retrieved from http://www.youtube.com/watch?v= HTrxgnAAcjA.

Hurlbut, R. (2009b, September 7). San Diego Trolley Police Prohibit Photography. *The World is Raw.com* Retrieved from http://www.theworldisraw.com/san-diego-trolley-police-prohibit-photography/.

Hurlbut, R. (2009c, September 21). San Diego Trolley Police Were Wrong. *YouTube.com*: Theworldisraw. Retrieved from http://www.youtube.com/watch?v=UVRqinCLXEU.

Hurlbut, R. (2009d, September 21). San Diego Trolley Police Were Wrong. *The World is Raw.com* Retrieved from http://www.theworldisraw.com/san-diego-trolley-police-were-wrong/.

Hurlbut, R. (2010a, February 25). San Diego Reader Story: We Don't Want You Taking Pictures. *The World is Raw.com* Retrieved from http://www.theworldisraw.com/sdreader-story/.

Hurlbut, R. (2010b). Illegal Photography. *The World is Raw.com* Retrieved from http://www.theworldisraw.com/illegal-photography/.

Hutton, A. (2007, June 4). Arrested youths demand better police treatment. *Make the Road New York*. Retrieved from http://www.maketheroad.org/article.php?ID=346.

Hutton, C. (2008, September 24). Deranged man fatally Tasered. *Fox News*. Slideshow. Retrieved from http://www.foxnews.com/photoessay/0,4644,5238,00.html/#1_0.

IBM. (n.d.). IBM 350 disk storage unit. IBM Archives. Retrieved from http://www-03.ibm.com/ibm/history/exhibits/storage/storage_350.html.

In re Eastman Kodak Co., 12-10202, (2012). U.S. Bankruptcy Court, Southern District of New York (Manhattan).

Ingersoll, E. (1878, July). The Police of New York. *The Century*, or *Scribner's Monthly*. *XVI*(3), 342-356.

Ingham, J. N. (1983). *Biographical Dictionary of American Business Leaders*, v. 1. Westport, CT: Greenwood Press.

Italiano, L. (2010a, April 23). Ex-cop's bike spin revealed. *New York Post*. Retrieved from http://www.nypost.com/p/news/local/manhattan/ex_cop_bike_spin_revealed_kiMiszf4IGBk2mQMDNPbiI.

Italiano, L. (2010b, July 14). Bike-push cop spared jail, probation in punishment-free sentencing. *New York Post*. Retrieved from http://www.nypost.com/p/news/local/manhattan/bike_shove_cop_gets_no_jail_time_RqH8DilfrnRwGWJSUm2o3O.

Jackson et al. v. City of New York et al., 1:09-cv-02715-SJ-SMG (2009, June 22). Document 1: Complaint against City of New York, Raymond Kelly, Scott Henderson, John H. Bambury, Louis G. Morselli, Edward Paszel, Dilson Abreu, Michael McCready, Jared E. Santangelo, Edwin LaPorte, Joseph Riso, John Does One through Twelve, filed by Merchant Trapp, Zezza Anderson, Joseph Griffin, Daniel Torres, Marvette Jackson, Leana Mejia, Angelina Castillo, Naquaisha Seaberry, Tina Jackson, Saviya Wilkerson, Asher Callender, Ezra Reneau. US District Court, Eastern District of New York. Retrieved from PACER.

Jackson et al. v. City of New York et al., 1:09-cv-02715-SJ-SMG (2010, March 2). Document 18: Settlement agreement. US District Court, Eastern District of New York. Retrieved from PACER.

Jackson, C. L. (2008a, October 27). Newark special officer suspended over CBS cameraman's arrest. *NJ.com*. Retrieved from http://www.nj.com/newark/index.ssf/2008/10/newark_special_officer_suspend.html.

Jackson, C. L. (2008b, October 27). Police ID two arrested in Newark fatal shootings. *NJ.com*. Retrieved from http://www.nj.com/newark/index.ssf/2008/10/police_id_two_arrested_in_newa.html.

Jackson, C. L. (2008c, October 27). A call to Newark churches. *NJ.com*. Retrieved from http://www.nj.com/newark/index.ssf/2008/10/a_call_to_newark_churches.html#more.

Janis, S. (2010, July 20). Skater hater — Trial board clears cop of most serious charges in skateboard incident. *Investigative Voice*. [blog]. Retrieved from http://www.investigativevoice.com/index.php?option=com_content&view=article&id=4994:news-flash-trial-board-clears-officer-salvatore-rivieri-of-most-serious-charges-related-to-skateboard-incident&catid=25:the-project&Itemid=44.

JaviErick. (2008, February 16). Baltimore cop strikes again. Retrieved from http://www.youtube.com/watch?v=l3Mts9gMGKo.

Jenkins, R. (1975). *Images and Enterprise: Technology and the American Photographic Industry, 1839 to 1925*. Johns Hopkins Studies in the History of Technology. Baltimore: Johns Hopkins University Press.

Jimster1956. (2009, September 28). Baltimore crime beat: Skateboarder versus cop. Comments. *The Baltimore Sun*. Retrieved from http://weblogs.baltimoresun.com/news/crime/blog/2009/09/skateboarder_versus_cop.html#comments.

Johnson et al. v. Bay Area Rapid Transit et al., No. CV 09-00901 MHP (2011, May 9). Document 207: Memorandum & order re: cross motions for summary judgment. Retrieved from LexisNexis Academic.

Johnson et al. v. Bay Area Rapid Transit et al., No. CV 09-00901 MHP (2011, November 22). Document 243: Order approving distribution and allocation of settlement funds. Retrieved from PACER.

Johnston, R. (1940). Slanguage of Amateur Photographers. *American Speech, 15*(4), 357–360.

Johnstone, B. (1999). We Were Burning: Japanese Entrepreneurs and the Forging of the Electronic Age. New York, NY: Basic Books.

Joint Photograph Experts Group. (1992). *ISO/IEC FDIS 10918-5*.

Jones, B., & Sentementes, G. G. (2008, February 14). Skateboarder calls reaction over the top. *The Baltimore Sun*. Retrieved from http://articles.baltimoresun.com/2008-02-14/news/0802140242_1_santo-bush-skateboarding.

Kanter, B. & Fine, A. (2010). The Networked Nonprofit: Connecting with Social Media to Drive Change. New York: Wiley.

Karr, J-B. A. (1849, January). *Les Guêpes*. In Manser, M. H. (Ed.), (2007). *The Facts on File Dictionary of Proverbs*. New York: Infobase. 191.

Kattelle, A. D. (1986). The Evolution of Amateur Motion Picture Equipment 1895-1965. *Journal of Film & Video, 38*(3/4), 47 - 57.

Kattelle, A. D. (2000). *Home Movies: A History of the American Industry, 1897-1979*. Nashua, NH: Transition Publishing.

Kearney, B. (2008, December 11). 'YouTube' skateboarder can sue cop, rules Baltimore City Circuit judge. *The Daily Record*, Baltimore MD. Retrieved from LexisNexis Academic.

Kellner, L. (2009, August 7). Bushwick 32 teach lesson in organized struggle. *Liberation News*. Retrieved from http://pslweb.org/liberationnews/news/09-08-07-bushwick-32-teach-lesson-in-orga.html.

Kelly, D. (2010a). *Watching YouTube: Extraordinary Videos by Ordinary People* [review of the book by M. Strangelove]. Unpublished manuscript, Department of Communication & Journalism, University of Maine, Orono, Maine.

Kelly, D. (2010b). *YouTube: Online Video and Participatory Culture* [review of the book by J. Burgess & J. Green]. Unpublished manuscript, Department of Communication & Journalism, University of Maine, Orono, Maine.

Kelly, D. A. (2010). *Internet Imagery and the Public Sphere*. Unpublished manuscript. Department of Communication & Journalism, University of Maine, Orono.

Kelly, D. A. (2011). *Police-civilian Interaction and Response to Civilian Photography, 1839-2010*. Unpublished manuscript. Department of Communication & Journalism, University of Maine, Orono.

Kelly, D. A. (2012). *Cops, cameras and accountability: User-generated online video and public space police-civilian interactions* (Ph.D.). The University of Maine, Ann Arbor.

Kennedy, J. (2010a). Don't you forget about me -- An exploration of the "Maddie Phenomenon" on YouTube. *Journalism Studies, 11*(2), 225 - 242. doi:10.1080/14616700903290635.

Khan, S. (2010, April 19). Tourist takes stand in first day of officer assault trial. *NY1 News*. Retrieved from http://manhattan.ny1.com/content/top_stories/117156/ tourist-takes-stand-in-first-day-of-officer-assault-trial.

KIRO-TV. (2010a, May 7). Watch it: Why was footage not shown earlier. Retrieved from http://www.kirotv.com/videos/news/watch-it-why-was-footage-not-shown-earlier/vCk6n/.

KIRO-TV. (2010b, May 7). Uncut: Freelance photographer speaks out. Retrieved from http://www.kirotv.com/videos/news/uncut-freelance-photographer-speaks-out/vCk6f/.

KIRO-TV. (2010c, May 7). Uncut: SPD officer statement. Retrieved from http://www.kirotv.com/videos/news/uncut-spd-officer-statement/vCk6p/.

KIRO-TV. (2010d, May 7). Seattle mayor issues statement about police video.

KIRO-TV. (2010e, May 10). Police attack video causes firestorm of controversy: Officers reassigned as police conduct investigation. Retrieved from http://www.kirotv.com/news/ news/police-attack-video-causes-firestorm-of-controvers/nDrS2/.

KIRO-TV. (2010f, May 10). Watch it: Latino organization says they're helping victim of police violence. Retrieved from http://www.kirotv.com/videos/news/watch-it-latino-organization-says-theyre-helping/vCk6z/.

KIRO-TV. (2010g, May 10). Watch it: Spanish radio stations buzzing with debate. Retrieved from http://www.kirotv.com/videos/news/watch-it-spanish-radio-stations-buzzing-with/vCk6x/.

KIRO-TV. (2010h, May 10). Watch it: Seattle police reports contradict video of attack. Retrieved from http://www.kirotv.com/videos/news/watch-it-seattle-police-reports-contradict-video/vCky7/.

KIRO-TV. (2010i, May 10). Female officer in cop stomping video identified. Retrieved from http://www.kirotv.com/news/news/female-officer-in-cop-stomping-video-identified/nDrS4/.

KIRO-TV. (2010j, May 11). FBI launches investigation into officer stomping video. Retrieved from http://www.kirotv.com/news/news/fbi-launches-investigation-into-officer-stomping-v/nDPS7/.

KIRO-TV. (2010k, May 11). Uncut: NAACP news conference. Retrieved from http://www.kirotv.com/videos/news/uncut-naacp-news-conference/vCk7B/.

KIRO-TV. (2010l, May 12). SPD officer previously investigated for police brutality. Retrieved from http://www.kirotv.com/news/news/spd-officer-previously-investigated-for-police-bru/nDrS5/.

KIRO-TV. (2010m, May 13). What happened before police stomping video. Retrieved from http://www.kirotv.com/news/news/what-happened-before-police-stomping-video/nDPSw/.

KIRO-TV. (2010n, May 13). Robbery victim describes crime that led to police attack. Retrieved from http://www.kirotv.com/news/news/robbery-victim-describes-crime-that-led-to-police-/nDrS6/.

KIRO-TV. (2010o, May 14). Diaz will stay in running for top cop job. Retrieved from http://www.kirotv.com/news/news/diaz-will-stay-in-running-for-top-cop-job/nDPS5/.

KIRO-TV. (2010p, May 14). Victim of police stomping issues statement. Retrieved from http://www.kirotv.com/news/news/victim-of-police-stomping-issues-statement/nDPMN/.

KIRO-TV. (2010q, May 18). Watch it: Investigation into if crimes were committed by officers in stomping incident. Retrieved from http://www.kirotv.com/videos/news/watch-it-investigation-into-if-crimes-were/vCm2n/.

KIRO-TV. (2010r, May 18). Latino leaders call for removal of 2 Seattle officers. Retrieved from http://www.kirotv.com/news/news/latino-leaders-call-for-removal-of-2-seattle-offic/nDPWQ/.

KIRO-TV. (2010s, May 19). Seattle officers now subject of criminal investigation. Retrieved from http://www.kirotv.com/news/news/seattle-officers-now-subject-of-criminal-investiga/nDrS8/.

KIRO-TV. (2010t, May 21). SPD opens probe over whether police tried to suppress video. Retrieved from http://www.kirotv.com/news/news/spd-opens-probe-over-whether-police-tried-to-suppr/nDrRz/.

KIRO-TV. (2010u, July 6). Active, ongoing criminal probe under way in Seattle cop stomp case. Retrieved from http://www.kirotv.com/news/news/active-ongoing-criminal-probe-under-way-in-seattle/nDRbq/.

KIRO-TV. (2010v, August 10). Videographer denies charges he stole TV gear. Retrieved from http://www.kirotv.com/news/news/videographer-denies-charges-he-stole-tv-gear/nDN8b/.

KIRO-TV. (2010w, September 1). No hate crime charges for officer in stomping incident. Retrieved from http://www.kirotv.com/news/news/no-hate-crime-charges-for-officer-in-stomping-inci/nDRms/.

KIRO-TV. (2010x, November 9). Seattle cop in stomping incident accused of abusing another suspect. Retrieved from http://www.kirotv.com/news/news/seattle-cop-in-stomping-incident-accused-of-abusin/nDSgg/.

KIRO-TV. (2011a, February 11). SPD officer from stomping video recorded flipping off suspect. Retrieved from http://www.kirotv.com/news/news/spd-officer-from-stomping-video-recorded-flipping-/nDmCm/.

KIRO-TV. (2011b, February 11). Watch it: Another embarrassment for Seattle officer. Retrieved from http://www.kirotv.com/videos/news/watch-it-another-embarrassment-for-seattle-officer/vCqjh/.

KIRO-TV. (2011c, April 26). Audio: Judge vents frustration, dismisses charge. Retrieved from http://www.kirotv.com/videos/news/audio-judge-vents-frustration-dismisses-charge/vCrSh/.

KIRO-TV. (2011d, April 26). Cop in stomping incident involved in police video cover-up. Retrieved from http://www.kirotv.com/news/news/cop-in-stomping-incident-involved-in-police-video-/nDmCh/.

KIRO-TV. (2011e, April 26). Watch it: More controversy for SPD officer. Retrieved from http://www.kirotv.com/videos/news/watch-it-more-controversy-for-spd-officer/vCrSj/.

KIRO-TV. (2011f, April 27). Seattle police: Cameras just weren't on during alleged choking incident. Retrieved from http://www.kirotv.com/news/news/seattle-police-cameras-just-werent-on-during-alleg/nDNH2/.

KIRO-TV. (2011g, April 27). Watch it: Each PD different in way cameras used. Retrieved from http://www.kirotv.com/videos/news/watch-it-each-pd-different-in-way-cameras-used/vCrSx/.

KIRO-TV. (2011h, May 12). Lawyer: City should pay stomping victim $750K or face lawsuit. Retrieved from http://www.kirotv.com/news/news/lawyer-city-should-pay-stomping-victim-750k-or-fac/nDNwM/.

KIRO-TV. (2011i, May 12). UNCUT: Chief Diaz's News Conference On Cobane. Retrieved from http://www.kirotv.com/videos/news/uncut-chief-diazs-news-conference-on-cobane/vCrY3/.

KIRO-TV. (2011j, May 12). UNCUT: Shandy Cobane reacts to punishment. Retrieved from http://www.kirotv.com/videos/news/uncut-shandy-cobane-reacts-to-punishment/vCrZB/.

KIRO-TV. (2011k, May 14). SPD officer in stomping incident will not be fired. Retrieved from http://www.kirotv.com/news/news/spd-officer-in-stomping-incident-will-not-be-fired/nDrRx/.

KIRO-TV. (2011l, June 22). Man in Seattle cop-stomping incident suing city, officers. Retrieved from http://www.kirotv.com/news/news/man-in-seattle-cop-stomping-incident-suing-city-of/nDPLy/.

KIRO-TV. (2011m, June 22). Uncut: Complete Monetti news conference. Retrieved from http://www.kirotv.com/videos/news/uncut-complete-monetti-news-conference/vCr95/.

KIRO-TV. (2011n, June 22). Uncut: Monetti's complete statement. Retrieved from http://www.kirotv.com/videos/news/uncut-monettis-complete-statement/vCr7Z/.

Klein, A. (2005, May 14). Reporter's unmarked SUV alarmed official. *The Washington Post*. Final edition, Metro section, B02. Retrieved from LexisNexis Academic.

Klein, A., and Wiggins, O. (2005, May 20). Johnson explains increase in crime; Shift in policing called unnecessary. *The Washington Post*. Friday final edition, Metro section, B01. Retrieved from LexisNexis Academic.

Klockars, C. B., Kutnjak Ivkovic, S., & Haberfeld, M. R. (2007). *Enhancing Police Integrity*. New York, NY: Springer.

Knapp, W., & New York (City). Commission to Investigate Alleged Police Corruption. (1972). *Commission Report. (With Summary and Principle Recommendations, issued August 3, 1972)*. New York: George Braziller.

Kolodner, M. (2008, September 28). NYPD seeks crisis intervention unit. *The New York Daily News*. Retrieved from http://www.nydailynews.com/news/2008/09/28/2008-09-28_nypd_seeks_crisis_intervention_unit-1.html.

Krages, B. P. (2006). The photographer's right: your rights and remedies when stopped or confronted for photography. Retrieved from http://www.krages.com/ThePhotographersRight.pdf.

Krages, B. P. (2007). Legal Handbook for Photographers : The Rights and Liabilities of Making Images (2nd ed.). Buffalo, NY: Amherst Media.

Kuhn, M. (2001). *The Land List*. Retrieved from http://www.rwhirled.com/landlist/ landhome.htm.

Kusch, F. (2004). Battleground Chicago: The Police and the 1968 Democratic National Convention. Westport, Conn.: Praeger.

Lambert v. Polk County, Iowa, 723 F. Supp. 128, 133 (S.D. Iowa 1989).

Lang, R. (2010, August 26). Father of teen in Rivieri video speaks out. *C4 Show*. WBAL, Baltimore MD.

La Vigne, N., Lachman, P., Matthews, A., & Neusteter, S. R. (Eds.). (2012). Key Issues in the Police Use of Pedestrian Stops and Searches: Discussion Papers From an Urban Institute Roundtable. Washington, DC: The Urban Institute. Retrieved from https://www.ncjrs.gov/App/Publications/abstract.aspx?ID=261596.

Le Beau, E. (2008, April 23). Faced with authority? Here's some advice. Sneakers, C6. *Herald News* (Passaic County, NJ).

Leavitt, L. (2009 October 12). The lieutenant's death that will not die. *NYPD Confidential*. Retrieved from http://nypdconfidential.com/columns/2009/091012.html.

Lee, T. (2007a, May 25). Apology is sought after arrests of 30 people on the way to a slain teenager's wake. *The New York Times*, p. B4. New York, NY. Retrieved from ProQuest Newsstand.

Lee, T. (2007b, June 24). In the details of a mass arrest, two versions, worlds apart. *The New York Times*, p. A.25. New York, NY. Retrieved from ProQuest Newsstand.

Lee, T. B. (2012a, January 10). Boston PD admits arrest for cell phone recording was a mistake. *Ars Technica*. Retrieved from http://arstechnica.com/tech-policy/news/2012/01/boston-pd-admits-arrest-for-cell-phone-recording-was-a-mistake.ars.

Lee, T. B. (2012b, January 12). Obama administration says Constitution protects cell phone recordings. *Ars Technica*. Retrieved from http://arstechnica.com/tech-policy/news/2012/01/obama-administration-says-constitution-protects-cell-phone-recordings.ars.

Lenhart, A., Ling, R., Campbell, S., & Purcell, K. (2010). *Teens and Mobile Phones*. Pew Internet & American Life Project (p. 114). Washington, D.C.: Pew Research Center. Retrieved from http://www.pewinternet.org/Reports/2010/Teens-and-Mobile-Phones.aspx.

Leonard, J. (2010, June 24). Dramatic video of BART shooting released by court. *Los Angeles Times*. Retrieved from http://latimesblogs.latimes.com/lanow/2010/06/dramatic-video-of-bart-shooting-released-by-court.html.

Lerman, A. E., & Weaver, V. (2014). Staying out of Sight? Concentrated Policing and Local Political Action. *The ANNALS of the American Academy of Political and Social Science, 651*(1), 202–219.

Levenson, L. R. (2001). Police corruption and new models for reform. *Suffolk University Law Review (35)*1, 18-25.

Levinson, P. (2004). Cell Phone: The Story of the World's Most Mobile Medium and How it has Transformed Everything! New York: Palgrave Macmillan.

LewAllen, D. (2010). Westland Police Department, officers named in lawsuit alleging abuse of a prisoner in holding cell. Detroit, MI: WXYZ. Retrieved from: http://www.wxyz.com/

dpp/news/region/wayne_county/westland-police-department,-officers-named-in-lawsuit-alleging-abuse-of-a-prisoner-in-holding-cell.

Linskey, A. (2008, April 23). Police shift Inner Harbor patrol command. *The Baltimore Sun.* B7.

Linskey, A., & Sentementes, G. G. (2008, February 12). Inner Harbor incident hits Internet: Officer suspended after video shows him berating teen. McClatchy - Tribune Business News. Retrieved from http://proquest.umi.com/pqdweb?did=1427518551.

Livingston, D. (1997). Police discretion and the quality of life in public places: Courts, communities, and the new policing. *Columbia Law Review, 97*(3), 551-672.

Livingston, D. (2004). The unfulfilled promise of citizen review. *Ohio State Journal of Criminal Law, 1*(653), 654-655.

Loevy, J. (2010, Spring). Truth or consequences - police "testilying". *Litigation (36)*3, 13-20.

Long v. The City of New York et al., 109-cv-06099-AKH (2009a, July 7) Document 1: Complaint. Retrieved from PACER.

Long v. The City of New York et al., 109-cv-06099-AKH (2009b, September 29) Document 9: Answer of Defendant City of New York. Retrieved from PACER.

Long v. The City of New York et al., 109-cv-06099-AKH (2009c, October 19) Document 12: Judgment pursuant to Rule 68. Retrieved from PACER.

Long v. The City of New York et al., 109-cv-06099-AKH (2010, December 1) Document 45: Stipulation and order of settlement and discontinuance of plaintiff's attorneys' fees and costs. Retrieved from PACER.

Long, C. (2008 October 3). New York officer kills himself over Taser episode. Associated Press Worldstream. Domestic News. Retrieved from LexisNexis Academic.

Long, C. (2009, February 25). Victim's family in NYPD stun gun death sues city. Associated Press State & Local Wire. State and Regional. Retrieved from LexisNexis Academic.

Los Angeles Times (losangelestimes). (2010, June 24). Court releases dramatic video of BART shooting. *YouTube*. Retrieved from http://www.youtube.com/watch?v=Q2LDw5l_yMI.

Madden, M. (2007). Online Video. *Pew Internet & American Life Project* (p. 28). Washington, D.C.: Pew Research Center. Retrieved from http://www.pewinternet.org/Reports/2007/ Online-Video.aspx.

Maddox, B. (1972). *Beyond Babel: New Directions in Communications.* New York: Simon and Schuster.

Maghan, J., O'Reilly, G. W., & Shon, P. C. H. (2002). Technology, Policing, and Implications of In-Car Videos. *Police Quarterly*, *5*(1), 25 -42. doi:10.1177/109861102129198002.

Magoun, A. B. (2009). *Television: The Life Story of a Technology*. Baltimore, MD: Johns Hopkins University Press.

Mann, S., Nolan, J., & Wellman, B. (2003). Sousveillance: Inventing and using wearable computing devices for data collection in surveillance environments. *Surveillance & Society 1*(3). 331-355.

Mapp v. Ohio, 367 U.S. 643 (1961).

Marcus, A. I., & Segal, H. P. (1999). *Technology in America: A Brief History* (2nd ed.). Fort Worth: Harcourt Brace College Publishers.

Martin, G. (2008). A picture is worth a thousand words. *The Phrase Finder*. Retrieved from http://www.phrases.org.uk/meanings/a-picture-is-worth-a-thousand-words.html.

Martin, M. D. (n.d.). Movie cameras. Retrieved April 16, 2011, from http://www.mrmartinweb.com/movie.html.

Marx, G. (2003). A tack in the shoe: Neutralizing and resisting the new surveillance. *Journal of Social Issues 59*(1).

Maryland General Assembly. (2011). House Bill 45: Fiscal and policy note. Criminal Law - Interception of Oral Communications - Law Enforcement Officers. Department of Legislative Services. Retrieved from http://mlis.state.md.us/2011rs/fnotes/bil_0005/ hb0045.pdf.

Maryland General Assembly. (2011 February 18). House Judiciary Committee Voting Record 2011 Session: HB45. Retrieved from http://mlis.state.md.us/2011rs/votes_comm/hb0045_jud.pdf.

Maryland Judiciary. (2011 August 16). Case information: 12-K-10-0647 State Of Maryland vs Anthony John Graber III. Circuit Court for Harford County - Criminal System.

Maryland Judiciary. (2011). Case information: CAL07-10564 MCCAREN VS PRINCES GEORGES. Circuit Court for Prince George's County - Civil System. Retrieved from http://casesearch.courts.state.md.us/inquiry/inquiryDetail.jis?caseId=CAL0710564&loc=65&detailLoc=PGV.

Maryland v. Graber, 12-K-10-0647 (2010a, July 1). Document 14: Memorandum of law in support of defendant's motions to suppress. Circuit Court of Maryland, Harford County.

Maryland v. Graber, 12-K-10-0647 (2010b, September 27). Document 37: Opinion and Order. Circuit Court of Maryland, Harford County.

Marzulli, J. (2010, July 10). Two cuffed Brooklyn women hit back at NYPD officers with wrongful arrest accusation. *New York Daily News*, p. 14. New York, NY. Retrieved from http://articles.nydailynews.com/2010-07-10/local/29440046_1_nypd-officers-women-claim-arrest.

MarzuqVision. (2008, October 28). Newark police strangle First Ammendment. *YouTube*. Retrieved from http://www.youtube.com/watch?v=CTAAZBT0rBU.

Masnick, M. (2005, January 5). Since when is it illegal to just mention a trademark online? *Techdirt*. Retrieved from http://www.techdirt.com/articles/20050105/ 0132239.shtml.

Massachusetts v. Glik, No. 0701 CR 6687 (2008, January 31). Order of dismissal. Retrieved from http://volokh.com/files/glik.pdf.

Massachusetts v. Glik, No. 0701 CR 6687, slip op. at 3-4 (Boston Mun. Ct. Jan. 31).

McCarren v. Prince George's County et al., CAL07-10564 (2007, April 16). Document 1: Complaint and jury demand.

McCarren v. Prince Georges County et al., CAL07-10564 (2009, January 28). Document 133: Civil judgment sheet.

McCarren, A. (2007, Summer). Words and reflections: Two years later, justice denied. *Nieman Reports*. 64. Retrieved from http://www.nieman.harvard.edu/reportsitem.aspx?id=100231.

McCombs, M. & Reynolds, A. (2009). How the news shapes our civic agenda. In J. Bryant & M. B. Oliver (Eds.), *Media Effects: Advances in Theory and Research* (3rd ed.) (pp. 1-16). New York: Routledge.

McDonald, R. N. (2010, July 7). Maryland Wiretap Act. Office of the Attorney General, State of Maryland. Retrieved from http://www.oag.state.md.us/Topics/ WIRETAP_ACT_ROSENBERG.pdf.

McGuire, M. (2010). Application for statement of charges. Retrieved from http://www.pixiq.com/sites/default/files/blog/carlosmiller/uploads/2010/04/AnthonyGraber_939x689.jpg.

McKnight, G. (2011, December 14). Re: File no. IIS 10-175. Los Angeles Police Department. Retrieved from http://seattletimes.nwsource.com/ABPub/2010/12/21/2013733506.pdf.

McNerthney, C. (2010a, May 6). Police video: Cameraman tells his side of race-related incident. seattlepi.com. Retrieved from http://www.seattlepi.com/local/article/Police-video-Cameraman-tells-his-side-of-898323.php.

McNerthney, C. (2010b, August 6). Photographer who captured controversial video arrested. KOMONews.com. Retrieved from http://www.komonews.com/news/local/ 100172149.html.

McNerthney, C. (2011, June 7). Seattle police management slams city attorney. seattlepi.com. Retrieved from http://www.seattlepi.com/local/seattle-history/article/Seattle-police-management-slams-City-Attorney-1414196.php.

McPherson, K. (2010, April 8). MD motorcycle speeder charged for posting traffic stop video. *WJZ-TV*. Retrieved from http://wjz.com/local/traffic.stop.youtube.2.1621320.html.

Meister, A. (2010a, July 20). Officer Salvatore Rivieri of the YouTube skateboard incident. *Baltimore Politics Examiner.*

Meister, A. (2010b, November 17). Bull Roast to benefit Salvatore Rivieri in Parkville tonight. The Politics Meister. *Charm City Current* (A *Baltimore Sun* website). Retrieved from: http://charmcitycurrent.com/meister/2010/11/17/bull-roast-to-benefit-salvatore-rivieri-in-parkville-tonight/.

Merriam, S. B. (1988). *Case Study Research in Education: A Qualitative Approach*. The Jossey-Bass Social and Behavioral Science Series. San Francisco: Jossey-Bass.

Messing, P. and Shahid, A. (2008, September 27). Taser cops to retrain. *The New York Post*. Sports, Late City Final. 11. Retrieved from LexisNexis Academic.

Messmer, E. (1994, August 22). BellSouth Cellular ships personal communicator. *Network World*. 13.

Metropolitan Transit System. (n.d.). Security awareness brochure. Retrieved from http://www.sdmts.com/Security/documents/SecurityAwarenessBrochure.pdf.

Mfuni, T. (2007a). Bushwick 32 challenge "racist" funeral arrests. *New York Amsterdam News*, *98*(29), 3. Retrieved from EBSCOhost.

Mfuni, T. (2007b). Bushwick 32 become Student Coalition Against Racial Profiling (SCARP). *New York Amsterdam News*, *98*(31), 17. Retrieved from EBSCOhost.

Mfuni, T. (2007c, July 26). Voices that must be heard: Accused Bushwick students form coalition against racial profiling. *New York Community Media Alliance*. Retrieved from http://www.indypressny.org/nycma/voices/281/news/news_2/.

Middleton, K., & Lee, W. E. (2010). *The Law of Public Communication, 2010 Update* (7th ed.). Boston: Pearson Allyn and Bacon.

Miles, M. B., & Huberman, A. M. (1994). *Qualitative Data Analysis: An Expanded Sourcebook* (2nd ed.). Thousand Oaks, CA: Sage Publications.

Miletich, S. (2011, July 14). Seattle police sergeant disciplined over ethnic-insult incident. *The Seattle Times*. Retrieved from http://seattletimes.nwsource.com/html/localnews/ 2015614636_cobane15m.html.

Miller v. Rivieri. (2008). Case number 24C08002405. Circuit Court for Baltimore City.

Miller, C. (2010a, February 5). Photographer receives apology after armed guards harass him for shooting video. *Pixiq*. Retrieved from http://www.pixiq.com/article/san-diego-photographer-receives-apology-after-armed-guards-harass-him-for-taking-photos.

Miller, C. (2010b, February 8). Why the video mode in the new digital SLRs is so important. *Pixiq*. Retrieved from http://www.pixiq.com/article/why-the-video-mode-in-the-new-digital-slrs-is-so-important.

Miller, C. (2010c, February 26). Security firm refuses to release report of excessive force incident caught on video. *Pixiq*. Retrieved from http://www.pixiq.com/article/california-security-firm-refuses-to-release-report-of-excessive-force-incident-caught-on-video.

Miller, C. (2010d, April 16). Motorcyclist jailed for 26 hours for videotaping gun-wielding cop. *Pixiq*. Retrieved from http://www.pixiq.com/article/maryland-motorcyclist-spends-26-hours-in-jail-on-wiretapping-charge-for-filming-cop-with-gun.

Miller, C. (2010e). Photography is not a crime. Retrieved from: http://www.pixiq.com/ contributors/carlosmiller.

Miller, C. (2011). Photography is not a crime. Blog. *Pixiq.com*. Retrieved from http://www.pixiq.com/contributors/248.

Miller, C. C. (2010, September 2). YouTube ads turn videos into revenue. *The New York Times*. Retrieved from: http://www.nytimes.com/2010/09/03/technology/03youtube.html.

Miller, C. C. (2010, September 2). YouTube ads turn videos into revenue. *The New York Times*. Retrieved from: http://www.nytimes.com/2010/09/03/technology/03youtube.html.

Miller, R. G. (2010). A cop, a kid with no dad and going too far. *Practically Speaking*.[blog]. Retrieved from http://allaboutpr.rgmcomms.com/2010_08_01_archive.html.

Miller, W. R. (1975). Police authority in London and New York City 1830-1870. *Journal of Social History, 8*(2), 81-101.

Mills, A. J., Durepos, G., & Wiebe, E. (Eds.). (2010). *Encyclopedia of Case Study Research* (Vols. 1-2). Los Angeles, CA: SAGE Publications.

Miss Tam-Tam. (2008, October 28). Newark Special Police officer suspended after arresting TV cameraman. *NewarkSpeaks.com.* [Online forum message]. Retrieved from http://www.newarkspeaks.com/forum/showthread.php?t=8773.

Mitchell, D. (2003). The Right to the City: Social Justice and the Fight for Public Space. New York, NY: Guilford Press.

Mobile Phone News. (1993, November 8). Bellsouth, IBM unveil personal communicator phone. Retrieved from http://findarticles.com/p/articles/mi_m3457/is_n43_v11/ai_14297997/.

Mollen, M. (1994). *Commission Report. Commission to Investigate Allegations of Police Corruption and the Anti-Corruption Procedures of the Police Department.* New York, NY: The Commission.

Monetti v. City of Seattle, et al., No. 2:11-cv-01041 (2011a, June 11). Document 1: Complaint. Retrieved from PACER.

Monetti v. City of Seattle, et al., No. 2:11-cv-01041 (2011b, July 18). Document 8: Answer to complaint with jury demand by City of Seattle, Shandy Cobane, Mary L Woollum. Retrieved from PACER.

Monetti v. City of Seattle, et al., No. 2:11-cv-01041 (2011c, August 8). Document 13: Order reassigning case to Judge Ricardo S. Martinez. Retrieved from PACER.

Monetti v. City of Seattle, et al., No. 2:11-cv-01041 (2011d, August 8). Document 14: Order setting trial date and related dates. Retrieved from PACER.

Monetti v. City of Seattle, et al., No. 2:11-cv-01041 (2012, June 22). Document 62: Order on defendant's 24 motion for summary judgment by Judge Ricardo S Martinez. Retrieved from http:// dockets.justia.com/docket/washington/wawdce/2:2011cv01041/176702.

monkeyassj. (2009, January 5). BART police shooting in Oakland KTVU report. *YouTube*. Retrieved from http://www.youtube.com/watch?v=IKy-WSZMklc.

Monkkonen, E. H. (1992). History of urban police. *Crime and Justice, 15*, Modern Policing, 547-580.

Morris, J. (stringer253) (2010, May 29). The Seattle Police beating by- Jud Morris the raw unedited and origanal version. *YouTube*. Retrieved from http://www.youtube.com/watch?v=ue0xyCBaT-4.

Moving Picture Experts Group. (1995). *ISO/IEC 13818.*

MoxNewsDotCom. (2010, May 7). Seattle cop's tearful apology for beating an innocent man. *YouTube*. Retrieved from https://www.youtube.com/watch?v=ZrHbNN27ccU.

mpizzie. (2009, April 21). bushwick 32. *YouTube*. Retrieved from http://www.youtube.com/ watch?v=roV_1YRv260.

Mulberry Street. (ca. 1890). [gelatin silver print]. (X2010.11.2622, MNY216893). Museum of the City of New York. 1220 Fifth Avenue, New York, NY. Retrieved from http://collections.mcny.org.

Nakamura, K. (2004). *Classic Cameras: Kodak MotorMatic*. Retrieved from http://www.photoethnography.com/ClassicCameras/KodakMotormatic.html.

Nalder, E. (2010, April 20). Local computer security expert investigates police practices. *seattlepi.com*. Retrieved from http://www.seattlepi.com/local/article/Local-computer-security-expert-investigates-887914.php.

Narayan, S. (2010, May 12). Racial profiling video. *ACLU-WA.org*. Retrieved from http://www.aclu-wa.org/blog/racial-profiling-video.

National Institute of Justice. (2003, July). CCTV: Constant cameras track violators. *NIJ Journal* (249). 16-23. Washington, DC: U.S. Department of Justice. Retrieved from: http://www.ncjrs.gov/pdffiles1/jr000249d.pdf.

National Press Club. (2005, May 13). National Press Club calls on Prince Georges County Police to release tapes. Press release. *U.S. Newswire*. Retrieved from ProQuest Newsstand.

National Public Radio. (2008, February 14). The Bryant Park Project. Retrieved from LexisNexis Academic.

Necessary Illusions Productions. (2002). The Rodney King trials. *Seeing is Believing: The Handicam Revolution*. Documentary. Retrieved from http://www.seeingisbelieving.ca/ handicam/king/.

Negron v. City of New York et al., 1:09-cv-00944-SLT-SMG (2009, March 6). Document 1: Notice of removal incorporating verified complaint. Retrieved from PACER.

Negron v. City of New York et al., 1:09-cv-00944-SLT-SMG (2010, November 10). Document 51: Notice of motion to dismiss by Susan Pigott. Retrieved from PACER.

Negron v. City of New York et al., 1:09-cv-00944-SLT-SMG (2010, December 20). Document 53: City defendants' motion to dismiss the complaint. Retrieved from PACER.

Negron v. City of New York et al., 1:09-cv-00944-SLT-SMG (2010, December 20). Document 54: Declaration of Rachel Seligman Weiss and supporting exhibits with notice of motion by Nicholas Marchesona, the city of New York. Retrieved from PACER.

Negron v. City of New York et al., 1:09-cv-00944-SLT-SMG (2010, December 20). Document 55: Memorandum in opposition filed by Olga Negron. Retrieved from PACER.

Negron v. City of New York et al., 1:09-cv-00944-SLT-SMG (2010, December 20). Document 56: Declaration of Alison Keena and supporting exhibits by Olga Negron. Retrieved from PACER.

Negron v. City of New York et al., 1:09-cv-00944-SLT-SMG (2010, December 20). Document 57: Reply in support filed by Nicholas Marchesona, the City of New York. Retrieved from PACER.

Nestel, T. J. (2006, March). Using surveillance camera systems to monitor public domains: Can abuse be prevented? M.A. thesis. Monterey, CA: Naval Postgraduate School. Retrieved from https://www.hsdl.org/homesec/docs/theses/06Mar_Nestel.pdf.

New York Civil Liberties Union, & American Civil Liberties Union of New York State. (2004). Palm card: What to do if you're stopped by the police (English and Spanish). *New York Civil Liberties Union.* Retrieved from http://www.nyclu.org/publications/palm-card-what-do-if-youre-stopped-police-english-and-spanish-2004.

New York Civil Liberties Union. (2006). *Who's watching? Video camera surveillance in New York City and the need for public oversight.* Retrieved from www.nyclu.org/pdfs/ surveillance_cams_report_121306.pdf.

New York Criminal Procedure Law, (n.d.). Article 170 - § 170.55, Adjournment in Contemplation of Dismissal. Retrieved from http://law.onecle.com/new-york/criminal-procedure/CPL0170.55_170.55.html.

New York Police Department. (2009, April 3). *Operations order 14: Investigation of individuals engaged in suspicious photography and video surveillance.* Retrieved from http://craphound.com/images/operationsorderinvestigationofindividuals engagedinsuspiciousphotographyandvideosurveillance.jpg.

New York Post. (2010, April 19). Cop and cyclist clash: a second video angle shown as evidence against police. Retrieved from http://www.nypost.com/p/news/local/ manhattan/push_cop_pogan_found_guilty_of_lying_JiSHQKqu2Vlm55K NLCwo2H.

New York Times Guide to Essential Knowledge: A Desk Reference for the Curious Mind. (2004). New York: St. Martin's Press.

New York Times. (1883, October 31). A lost camera. *The New York Times (1857-1922)*, 4. Retrieved from ProQuest Historical Newspapers.

New York Times. (1908, March 29). Bomb kills one; police escape. *The New York Times (1857-1922)*, 1. Retrieved from ProQuest Historical Newspapers.

New York Times. (1981, September 25). New scene stealer: The video camera. Late Edition (East Coast), p. D1.

New York v. Long. 896602. (2008, July 25). Statement of Patrick Pogan. Criminal Court of the City of New York, County of New York. Retrieved from http://animalnewyork.com/wp-content/uploads/archive/patrick-pogan-nypd.pdf.

Newman, A. (2008, January 29). Charges dropped for 22 arrested on way to wake. *The New York Times*, p. B3. New York, NY. Retrieved from ProQuest Newsstand.

Nielsen Company. (2010, October). Total U.S. Market & Smartphone Market. *Mobile Insights, National*. Retrieved from http://www.penipress.com/wp-content/uploads/2011/04/Screen-shot-2011-04-10-at-5.42.28-PM.png.

ninja636b. (2007, February 25). Fort Lauderdale – arrest. YouTube. Retrieved from http://www.youtube.com/watch?v=s3HrKY3abz8.

Noble, A. (2010, April 13). FBI to investigate student beatings at UM. *Gazette.net*. Retrieved from http://ww2.gazette.net/stories/04132010/prinnew153805_32578.php.

Nokia. (2005, April 27). *Twist and shoot: Nokia N90 makes its première*. Retrieved from http://www.prnewswire.com/news-releases/twist-and-shoot-nokia-n90-makes-its-premiere-54418187.html.

Nokia. (2006, September 26). *It's what computers have become – the new Nokia N95*. Press release. Retrieved from http://www.nokia.com/NOKIA_COM_1/Press/Press_Events/ Nokia_Open_Studio_26_Sept_2006/pdf/Nokia_N95_Press_Release.pdf.

Nokia. (2010). *N8-00 user guide*. Retrieved from http://nds1.nokia.com/phones/files/ guides/Nokia_N8-00_UG_en.pdf.

Nott, J. F. (1891, June 6). Photography and illustrated journalism. In *Wilson's Photographic Magazine 28*(395) 321-325.

Nuckols, B. (2010 September 27). Wiretap charges tossed in videotaping of trooper. Associated Press State & Local Wire. Retrieved from LexisNexis Academic.

NY1 News. (2008a, July 28). Online video shows city policeman pushing cyclist in protest. Retrieved from http://www.ny1.com/content/top_stories/84281/online-video-shows-city-policeman-pushing-cyclist-in-protest.

NY1 News. (2008b, September 24). Nude man reportedly Tased to death in Brooklyn. Retrieved from http://www.ny1.com/content/top_stories/86260/nude-man-reportedly-tased-to-death-in-brooklyn/.

NY1 News. (2010, October 19). City reaches settlement with Critical Mass bike riders. Retrieved from http://www.ny1.com/content/top_stories/127356/city-reaches-settlement-with-critical-mass-bike-riders.

NYPD stop-and-frisk: Massive injustice against the people that must be stopped. (2010, September 26). *Revolution*. Chicago, IL. Retrieved from http://revcom.us/a/212/stop_n_frisk-en.html.

O'Connor, J. J. (1977, January 2). Documentary on police strips away any glamour. *The New York Times*, 73.

O'Connor, J. J. (1988, April 12). Review/television; Video verite style for police story. *The New York Times*, Late Edition (East Coast) C18.

O'Hagan, M. (2010a, May 7). TV station passed up video, freelancer says. *Seattle Times*. Retrieved from http://seattletimes.nwsource.com/html/localnews/2011812020_video08m.html.

O'Hagan, M. (2010b, May 14). TV news director quits over videotape. *Seattle Times*. Retrieved from http://seattletimes.nwsource.com/html/localnews/2011860859_kcpq14m.html.

o1OpTiMuS1o. (2010, April 12). Video backs Maryland students police brutality claim 4-12-10. *YouTube*. Retrieved from http://www.youtube.com/watch?v=at1q3lIpK3c.

Officer.com. (2008, November 14). San Diego Trolley. *Forum*. Retrieved from http://forums.officer.com/forums/showthread.php?105147-San-Diego-Trolley.

Officer.com. (2010). Officer Rivieri. Public Forums: General Law Enforcement Topics. Retrieved from http://forums.officer.com/forums/showthread.php?152208-Officer-Rivieri.

OMB. (2005). *Approved Operating Budget FY2006*. Prince Georges County, MD. Retrieved from http://www.princegeorgescountymd.gov/Government/AgencyIndex/OMB/

approved_budget_fy2006/ApprovedOpBudget_FY2006/05_General-Gov/OMB_ApprOperating_FY06_General-Gov_46-52_CCOP.pdf.

Ott, C. (2010, February 1). ACLU challenges arrest for use of cell phone to capture police misconduct. *ACLU of Massachusetts*. Retrieved from http://aclum.org/news_2.1.10.

Ott, C. (2012, March 27). City of Boston pays $170,000 to settle landmark case involving man arrested for recording police with cell phone. *ACLU of Massachusetts*. Retrieved from http://aclum.org/news_3.27.12.

PACER. (2011a April 11). History/documents query with docket text: 11-17440 Anthony John Graber III, Chapter 7 Bankruptcy.

PACER. (2011b July 30). History/documents query with docket text: 1:10-cv-03067-SJ-RML *Chapman et al. v. City of New York et al.*

PACER. (2011c August 19). History/documents query with docket text: 2:09-cv-05396-FSH-PS *Quodomine v. City of Newark et al.*

PACER. (2011d October 15). History/documents query with docket text: 1:09-cv-00944-SLT-SMG *Negron v. The City of New York et al.*

PACER. (2011e December 18). History/documents query with docket text: 1:08-cv-02240-ENV-RLM *Green, Jr. et al. v. The City of New York et al.*

PACER. (2011f December 18). History/documents query with docket text: 1:08-cv-03034-ENV-RLM *Chavarria et al. v. City of New York et al.*

PACER. (2011g December 18). History/documents query with docket text: 1:08-cv-03269-ENV-RLM *Prosper et al. v. City of New York et al.*

PACER. (2011h December 18). History/documents query with docket text: 1:09-cv-02715-SJ-SMG *Jackson et al v. City of New York et al.*

PACER. (2011i December 29). History/documents query with docket text: 1:09-cv-06099-AKH *Long v. The City of New York et al.*

PACER. (2011j December 30). History/documents query with docket text: 3:09-CV-00901-MHP *Johnson et al v BART et al.*

PACER. (2012a January 1). History/documents query with docket text: 2:11-cv-01041-RSM *Monetti v. City of Seattle, et al.*

PACER. (2012b January 3). History/documents query with docket text: 10-1764 *Glik v. Cunniffe et al.* appeal from 1:10-cv-10150-WGY.

PACER. (2012c April 20). History/documents query with docket text: 1:10-cv-10150-WGY *Glik v. Cunniffe et al.*

Panasonic. (2004, April 18). *Panasonic AJ-SPX800 Camputer™ DVCPRO P2 solid-state memory camcorder delivering in May*. Press release. Las Vegas, NV: Panasonic. Retrieved from http://www.panasonic.com/business/provideo/news/news04_nab_11.asp.

Parascandola, R. (2008, October 3). Officer in Taser death who committed suicide probably would not have faced criminal charges. *Newsday*. Melville, NY. Domestic. Retrieved from LexisNexis Academic.

Parks, B. (2000, October). The Big Picture. *Wired, 8*(10). Retrieved from http://www.wired.com/wired/archive/8.10/kahn.html.

Patton, A. L. (1993). The endless cycle of abuse: Why 42 U.S.C. § 1983 is ineffective in deterring police brutality. *Hastings Law Journal* (44)753, 759-799.

Pavsner, S. (2007, April 16). Journalist sues Prince George's County for violation of constitutional rights. Press release. PRNewswire. Retrieved from http://www.prnewswire.com/news-releases/journalist-sues-prince-georges-county-for-violation-of-constitutional-rights-58331447.html.

Pearson, E., and Gendar, A. (2009, January 13). Taser victim's family furious as NYPD promotes zap cop. *The New York Daily News*. Retrieved from http://www.nydailynews.com/news/taser-victim-family-furious-nypd-promotes-zap-article-1.425001

Peltz, J. & Long, C. (2010, April 19). Ex-NYPD officer goes on trial over cyclist clash. Associated Press. Retrieved from LexisNexis Academic News.

Pennsylvania Prison Society, a.k.a. Philadelphia Society for Alleviating the Miseries of Public Prisons. (1858, January). Daguerreotype gallery of rogues. *The Journal of Prison Discipline and Philanthropy*. *XIII*(1), 45. Philadelphia: Edward C. and John Biddle.

Perry v. Los Angeles Police Dep't, 121 F.3d 1365, 1369 (9th Cir. 1997).

Perton, M. (2006). Samsung's new camcorders: SC-D165, SC-HDX15, SC-X210L and SC-X205L. *Engadget*. Retrieved from http://www.engadget.com/2006/01/12/ samsungs-new-camcorders-sc-d165-sc-hdx15-sc-x210l-and-sc-x20/.

Pescovitz, D. (2008 2). Cop roughs up teenage skateboarder on video - Boing Boing. Retrieved July 20, 2010, from http://boingboing.net/2008/02/13/cop-roughs-up-teenag.html.

Phantoscope Manufacturing Company (1912). Phantoscope projector advertisement. Retrieved from http://members.chello.nl/a.wichmann/phantascopeproj.jpg.

Piggott v. City of New York et al., NY Slip Op 51212(U) (2011, June 15). City defendants' cross motion to dismiss granted. Retrieved from http://law.justia.com/cases/new-york/other-courts/2011/2011-51212.html [Note misspelling of Pigott].

Pingdom. (2008, April 11). Royal Pingdom: Map of all Google data center locations. *Pingdom*. Retrieved from http://royal.pingdom.com/2008/04/11/map-of-all-google-data-center-locations/.

Plato. (ca. 380 B.C.E.). *The Republic*. Benjamin Jowett (Trans.). Project Gutenberg. Retrieved from: http://www.gutenberg.org/files/1497/1497-h/1497-h.htm.

Police in 3 states test, adopt Taser wearable camera. (2011, May 2). *Government Video*. Retrieved from http://www.governmentvideo.com/article/105732.

Police officers at Labor Day parade. (1890). [1 negative: glass; 4 x 5.25 in.]. Burton Historical Collection (EB02d187, BHC glass neg. no. 502). Detroit Public Library, Detroit, MI. Retrieved from http://quod.lib.umich.edu/cgi/i/image/image-idx?id=S-DPA1IC-X-EB02D187%5DEB02D187.TIF.

policeguide.com. (2011). New York Police Badges. J. Casey, Ed. Retrieved from: http://www.policeguide.com/Police_Photo_Galleries/Badge_Gallery_/New_York_Police__Badges/new_york_police__badges.html.

PoliceLink (2011, March 1). Video of officer's actions played major role in firing. Law enforcement news. Retrieved from http://policelink.monster.com/news/articles/ 151225-video-of-officers-actions-played-major-role-in-firing.

Policeman arresting a hobo on Mulberry Street. (ca. 1890). Bettmann Collection. New York, NY: Corbis. Retrieved from http://www.corbisimages.com.

Popular Science. (1949, February). Television Rides Wires. 179.

Potere, M. (2012). Who Will Watch the Watchmen?: Citizens Recording Police Conduct. *Northwestern University Law Review*, *106*(1), 273–316.

Potter, D. (2010). The YouTube Factor. *American Journalism Review, 32*(1), 46.

Present, B. (2010a, April 15). Riot place, riot time. *The Diamondback*. Retrieved from http://www.diamondbackonline.com/news/right-place-riot-time-1.1343680.

Present, B. (2010b, April 22). University Police locate missing riot video. *The Diamondback*. Retrieved from http://www.diamondbackonline.com/news/university-police-locate-missing-riot-video-1.1377637.

Present, B. (2010c, April 29). More student riot charges dropped. *The Diamondback.* Retrieved from http://www.diamondbackonline.com/news/more-student-riot-charges-dropped-1.1434784.

Prior, L. (2003). *Using Documents in Social Research. Introducing Qualitative Methods.* London; Thousand Oaks, Calif.: Sage Publications.

Prosper et al v. City of New York et al., 1:08-cv-03269-ENV-RLM (2008, August 14). Document 1: Complaint against Raymond Kelly, Scott Henderson, Louis G. Morselli, Edward Paszel, Dilson Abreu, Michael McCready, Jared E. Santangelo, John Does One through Twelve, City of New York, filed by William Prosper, Malik Farmer, Kenneth Frederick, Felix Vasquez, Daquan Walker, Jose Rodriguez, Ashante Callender. US District Court, Eastern District of New York. Retrieved from PACER.

Prosper et al v. City of New York et al., 1:08-cv-03269-ENV-RLM (2009, May 12). Document 18: Settlement agreement. US District Court, Eastern District of New York. Retrieved from PACER.

Prosser, W. L. (1960). Privacy. *California Law Review, 48*(3), 383-423.

Purcell, K. (2010). The state of online video. *Pew Internet & American Life Project* (p. 16). Washington, D.C.: Pew Research Center. Retrieved from http://www.pewinternet.org/ Reports/2010/State-of-Online-Video.aspx.

Purcell, K. (2013, October 10). Online Video 2013. Pew Research Center. Retrieved from http://pewinternet.org/Reports/2013/ Online-video.

Q13 Fox News. (2010, May 7). Officer's emotional apology. Retrieved from https://www.youtube.com/watch?v=ZrHbNN27ccU.

Q13 Fox News. (2011, September 9). Jud Morris pleads guilty to stealing Q13 FOX cameras: Placed in Veteran's Drug Court. Retrieved from http://www.q13fox.com/news/kcpq-jud-morris-pleads-guilty-to-stealing-q13-fox-cameras-20110909,0,1480573.story.

Quirke, C. (2008). Reframing Chicago's Memorial Day Massacre, May 30, 1937. *American Quarterly, 60*(1), 129-157. doi:10.1353/aq.2008.0009.

Quodomine v. City of Newark et al., 2:09-cv-05396-FSH-PS (2009, October 22). Document 1: Complaint against City of Newark, City of Newark Police Department, City of Newark Prosecutor's Office, Brian Sharif, Marvin Adames, Anna Pereira, John Does 1-10, filed by James Quodomine. Retrieved from PACER.

Quodomine v. City of Newark et al., 2:09-cv-05396-FSH-PS (2011, May 12). Document 86: Order dismissing case as settled, w/out costs & w/out prejudice. Retrieved from PACER.

Radio News Bureau. (1954). Television factbook. *Television Digest*, (50), 334.

Rainie, H., & Pew Internet & American Life Project. (2008). *Video sharing websites*. Retrieved from http://www.pewinternet.org/~/media/Files/Reports/2008/ Pew_Videosharing_memo_Jan08.pdf.pdf.

Reiss, A. J. (1971). *The Police and the Public*. New Haven, CT: Yale.

Reporters Committee for Freedom of the Press. (2006). A Reporter's Field Guide. *News Media & the Law*, 30(1), 1 - 8.

Reporters Committee for Freedom of the Press. (2007a). A primer on invasion of privacy. *News Media & the Law, 31*(4), 2 - 3. doi:AN 27696291.

Reporters Committee for Freedom of the Press. (2007b). State-by-State Guide to Privacy Law. *News Media & the Law*, 31(4), 4 - 16.

Reporters Committee for Freedom of the Press. (2008). Can We Tape? The Reporters Committee for Freedom of the Press. Retrieved April 5, 2010, from http://www.rcfp.org/taping/.

Reporters Committee for Freedom of the Press. (2010). The Reporters Committee for Freedom of the Press. Retrieved from http://www.rcfp.org/.

RepublicanRanting. (2008, October 27). Cop puts cameraman in chokehold for filming peaceful protest. *YouTube*. Retrieved from http://www.youtube.com/watch?v=LSsGrIf8F0M.

Reynolds, A. (2010, December 2). Housing Authority tenants want cameras, more security. *The Villager, 80*(27).

Rice, S. B. (1920). The Right of Privacy. University of Pennsylvania Law Review and American Law Register, 68(3), 284–287.

Richelson, J. T. (1999). U.S. Satellite Imagery, 1960-1999. *National Security Archive Electronic Briefing Book No. 13*. The National Security Archive, George Washington University. Retrieved from http://www.gwu.edu/~nsarchiv/NSAEBB/NSAEBB13/index.html.

Riendeau, D. (2010, June 8). Major news about the Glik case! *Mass Rights Blog: News and Comment about the ACLU of Massachusetts and civil liberties issues*. Retrieved from http://www.massrightsblog.org/2010/06/major-news-about-glik-case.html.

Riis, J. A. (1901). *The Making of an American*. New York: Macmillan.

Riis, J. A., & Yochelson, B. (2001). *Jacob Riis*. London: Phaidon.

Rivera, R., Baker, A., & Roberts, J. (2010, July 12). A few blocks, 4 years, 52,000 police stops. *The New York Times*, p. A1. New York, NY.

Roberts & Wood (Robertsandwood). (2010, December 10). Beating and arrest of Jack McKenna - April 03 2011. *YouTube*. Retrieved from http://www.youtube.com/watch?v=Zcrnnmt8cg8.

Roche, X. (January 2010). HTTrack website copier (version 3.43-9C) [computer program]. Retrieved from http://www.httrack.com/.

Romer, D., Jamieson, K. H., and de Coteau, N. J. (1998, June). The treatment of persons of color in local television news: ethnic blame discourse or realistic group conflict? *Communication Research, 25*(3) 286.

Rose, J. (2011, May 13). This is the police: Put down your camera. *Morning Edition.* National Public Radio. Retrieved from http://www.npr.org/2011/05/13/136171366/this-is-the-police-put-down-your-camera.

Rosenthal, (2011, December 23). Audit: Seattle police 'uneven' in using patrol-car cameras. *The Seattle Times*. Retrieved from http://seattletimes.nwsource.com/html/localnews/2017083900_spdvideo24m.html.

Rothman, C. (2008, October 27). Newark suspends special police officer after arrest of TV cameraman. *Newark Star-Ledger*. Newark, NJ. Retrieved from http://www.nj.com/news/index.ssf/2008/10/newark_special_police_officer.html.

Rowinski, D. (2010, January 12). Police fight cellphone recordings. New England Center For Investigative Reporting. Retrieved from http://www.boston.com/news/local/massachusetts/articles/2010/01/12/police_fight_cellphone_recordings/.

Rudovsky, D. (1992). Police abuse: Can the violence be contained?. *HARV. C.R.-C.L. L. REV.(27)*465, 481 n.60.

Ryan, J. (2009, November 3). Cameraman files suit against Newark officer who tried to halt filming of demonstration. *The Star-Ledger*. Newark, NJ. Retrieved from http://www.nj.com/news/index.ssf/2009/11/camerman_files_suit_against_ne.html.

Sadler, R. L. (2005). *Electronic Media Law*. Thousand Oaks, CA: Sage Publications.

Sager, M. (2004). Scary Monsters and Super Freaks: Stories of Sex, Drugs, Rock "n" Roll, and Murder. New York, NY: Thunder's Mouth Press.

San Diego Mass Transit System. (2007). *No Smoking Rider Alert*. PDF file. Retrieved from http://www.sdmts.com/marketing/documents/NoSmokingRiderAlert.pdf.

San Diego Trolley, Inc. (2006). Standard Services Agreement. Retrieved from http://www.theworldisraw.com/pics/sandiegotrolleypolicecontract.pdf.

Sanders, S. (2011). Samsung ITCAM-7. *Steve's Digicams*. Retrieved from http://www.steves-digicams.com/news/samsung-itcam-7---the-worlds-first-hdd-digital-camcorder.html.

Santo, T. (camb0i). (2008, February 9). Baltimore cops V.S. skateboarder. Retrieved from http://www.youtube.com/watch?v=9GgWrV8TcUc.

Santo, T. (camb0i). (2010). camb0i's channel. Retrieved from http://www.youtube.com/user/camb0i.

Sarnoff, D. (1968). *Looking Ahead: The Papers of David Sarnoff.* New York: McGraw-Hill.

Sasson, S. J. (2009). 1975 Kodak Prototype Digital Camera. *Camera: A history of Photography from Daguerreotype to Digital* (pp. 338-339). New York: Sterling Innovation.

Savov, V. (2010, October 14). Nokia N8 review. *Engadget*. Retrieved from http://www.engadget.com/2010/10/14/nokia-n8-review/.

Saxena, J. (2010, July 10). Women accuse NYPD of wrongful arrest. *Gothamist*. Retrieved from http://gothamist.com/2010/07/10/women_accuse_nypd_of_wrongful_arres .php.

Sayeed, N. (2010, April 14). Officer suspended after beating of UMd. student. *Associated Press Online*. Retrieved from LexisNexis Academic News.

Schnell v. City of Chicago, 407 F.2d 1084 (7th Cir. 1969).

Schrems, S. H. (2009). The making of an outlaw image: Al Jennings of Oklahoma. *Western Americana: History of the American West*. Norman, Oklahoma. Retrieved from http:// westernamericana2.blogspot.com/2009/12/making-of-outlaw-image-al-jennings-of.html.

Schultz, J. (2009). *No One was Killed: the Democratic National Convention, August 1968*. Chicago; London: University of Chicago Press.

Scott v. Harris, 550 U.S. 372 (2007). Retrieved from http://www.supremecourt.gov/ opinions/06pdf/05-1631.pdf.

Scott, T. (2010, August 26). Release of video prompts call for investigation: Chief deputy wants to know how traffic stop video was placed on web site. *Ledger-Sentinel*. Oswego, IL. Retrieved from: http://www.ledgersentinel.com/article.asp?a=9045.

Segraves, M. (2007, April 16). Local Reporter Sues Police Over Injuries. *WTOP Radio*. Retrieved from http://www.wtopnews.com/?nid=428&sid=1116072.

Segraves, M. (2009, January 16). Jury: Cops used excessive force against reporter. *WTOP Radio*. Retrieved from http://www.wtopnews.com/?nid=708&sid=1577281.

Semel, M. (2010, April 12). Video of confrontation after U-Md. basketball game shows police beating. *The Crime Scene* (blog). *The Washington Post*. Retrieved from http://voices.washingtonpost.com/crime-scene/video-of-confrontation-after-u.html.

Shapiro, D. (2008, September 24). New05c.jpg. Photograph. Retrieved from http://lookathimnow.files.wordpress.com/2008/09/new05c.jpg.

Sharp v. Baltimore City Police Department, et al., 1:11-cv-02888-BEL (2012, January 10) Document 24: Statement of Interest of the United States. United States District Court for the District of Maryland. Retrieved from http://www.justice.gov/crt/about/spl/ documents/Sharp_SOI_1-10-12.pdf.

Sherman, L. W. (1978). *Scandal and Reform: Controlling Police Corruption.* Berkeley: University of California Press.

Shin, A. (2010, June 16). From YouTube to your local court; Video of traffic stop sparks debate on whether police are twisting Md. wiretap laws. *The Washington Post*, p. A.1. Washington, DC. Retrieved from ProQuest Newsstand.

Siegler, M. G. (2010, November 21). In the era of the connected camera, the point & shoot commits seppuku. *TechCrunch*. Retrieved from http://techcrunch.com/2010/11/21/the-connected-camera/.

Simmons, K. C. (2010). New governance and the "new paradigm" of police accountability: A democratic approach to police reform. *Catholic University Law Review, 59*(Winter), 373-426.

Skehill, L. A. (2009). Cloaking police misconduct in privacy: Why the Massachusetts anti-wiretapping statute should allow for the surreptitious recording of police officers. *Suffolk University Law Review, 42*, 981-1012.

Skinner, J. P. (2008, November 7). Rivieri suspension hearing internal memo. Retrieved from http://weblogs.baltimoresun.com/news/crime/blog/2010/08/ fired_officer_in_skakeboard_in.html and from http://www.scribd.com/doc/36502335/Fax-000000436.

Skolnick, J. H. (1982). Deception by police. *Criminal Justice Ethics, 1*, 40-54.

Skolnick, J. H. (1994). Justice Without Trial: Law Enforcement in Democratic Society. New York, NY: Macmillan.

SMECC. (2007a). Sony CV series video. *Southwest Museum of Engineering, Communications and Computation*. Retrieved from http://www.smecc.org/sony_cv_series_video.htm.

SMECC. (2007b). Hitachi. *Southwest Museum of Engineering, Communications and Computation*. Retrieved from http://www.smecc.org/hitachi.htm.

Smith v. City of Cummings, 212 F.3d 1332, 1333 (11th Cir. 2000).

Smith, A. (2010). *Mobile Access 2010*. Pew Internet & American Life Project (p. 32). Washington, D.C.: Pew Research Center. Retrieved from http://www.pewinternet.org/ Reports/2010/Mobile-Access-2010.aspx.

Smith, G. (1998, October 12). Silicon eyes. *BusinessWeek*, (3599), 94.

Smith, I. (2011, April 5). Cost of hard drive space. *Nova Scotia's Electric Gleaner*. Canning, Nova Scotia. Retrieved from http://ns1758.ca/winch/winchest.html.

Snyder, K. (2010, February 24). We Don't Want You Taking Pictures. *San Diego Reader* (San Diego, CA). Retrieved from http://www.sandiegoreader.com/news/2010/feb/24/feature-we-dont-want-you-taking-pictures/.

Sontag, S. (1961). Spiritual style in the films of Robert Bresson. In *Against Interpretation, and Other Essays*. New York: Farrar, Straus and Giroux.

Sony. (2005, March). Company of firsts. Retrieved from http://news.sel.sony.com/en/ corporate_information/company_of_firsts.

Sony. (2010). Sony's Handycam camcorder celebrates its 25th anniversary. *Sony Insider*. Retrieved from http://www.sonyinsider.com/2010/05/24/sonys-handycam-camcorder-celebrates-its-25th-anniversary/.

Sprague, R. (2008). Orwell was an optimist: The evolution of privacy in the United States and its de-evolution for American employees. *The John Marshall Law Review*, 42, 83-134.

Springer v. Prince George's County et al., (2009). Complaint. Retrieved from http://ww2.gazette.net/links/files/springercomplaint.pdf.

Stafford-Fraser, Q. (1995). The Trojan Room Coffee Pot: A (non-technical) biography. Cambridge University Computer Laboratory. Retrieved from http://www.cl.cam.ac.uk/coffee/qsf/coffee.html.

Stanczak, G. C. (Ed.). (2007). *Visual research methods: image, society, and representation*. Los Angeles: SAGE Publications.

Stanley, J. (2013, October 9). Police Body-Mounted Cameras: With Right Policies in Place, a Win For All. ACLU.org. Retrieved from https://www.aclu.org/technology-and-liberty/police-body-mounted-cameras-right-policies-place-win-all.

Stannard, M. B. & Bulwa, D. (2009, January 7). BART shooting captured on video. *San Francisco Chronicle*. Retrieved from http://www.sfgate.com/cgi-bin/article.cgi?f=/c/a/2009/01/07/MNOV154P0R.DTL&ao=all.

Stark, S. (1987, September). Special issues and topics: Perry Mason meets Sonny Crockett: The history of lawyers and the police as television heroes. *University of Miami Law Review, 42,* 229.

Stark, S. (1997). *Dragnet* and the policeman as hero. In Glued to the Set: The 60 Television Shows and Events that Made Us Who We are Today (pp. 31-37). New York: Free Press.

State of Florida v. Rodriguez, Case No. 06015503CF10A. (2006, September 4). Case summary. Retrieved from http://www.clerk-17th-flcourts.org/Clerkwebsite/BCCOC2/OdysseyPA/CaseSummary.aspx?CaseID=287991 9&hidSearchType=party_case&DisplayCitation=yes&CaseNumber=06015503CF10A.

Steinberg, A. (1989). *The Transformation of Criminal Justice, Philadelphia, 1800-1880.* Chapel Hill: University of North Carolina Press.

Stevens, C. (2010, February 1). Suit filed over state's wiretap statue. *News Makers*. WBZ. Retrieved from http://www.podfeed.net/episode/Suit+filed+over+states+wiretap+statue/2290355.

Stowell, D. O. (2008). *The Great Strikes of 1877.* Urbana: University of Illinois Press.

Strangelove, M. (2010 May 17). YouTube in review. The Mark. Retrieved from: http://www.themarknews.com/articles/1499-youtube-in-review.

Strangelove, M. (2010). *Watching YouTube: Extraordinary Videos by Ordinary People*. Digital Futures Series. Toronto; Buffalo, NY: University of Toronto Press.

Stross, R. (2013, April 6). Wearing a Badge, and a Video Camera. NYTimes.com. Retrieved from http://www.nytimes.com/2013/04/07/business/wearable-video-cameras-for-police-officers.html.

Student Coalition Against Racial Profiling. (2009, June 12). *S.C.A.R.P.* Retrieved from http://ourdemandforjustice.org/site/.

Suptic, S. (2009). The TV camera: Past, present and future. (cover story). *Broadcast Engineering, 51*(3), 66-86.

Sykes, R. E., & Brent, E. E. (1983). *Policing, a Social Behaviorist Perspective.* Crime, law, and deviance series. New Brunswick, NJ: Rutgers University Press.

Szaniszlo, M. (2008, January 26). Lawyer put on trial for recording drug arrest. *Boston Herald.* News. 6. Retrieved from http://www.bostonherald.com/news/regional/view.bg?articleid=1069080&srvc=next_article.

Taft, R. (1938). Photography and the American Scene, a Social History, 1839-1889. New York: The Macmillan Company.

Tang, C. W., & VanSlyke, S.A. (1987). Organic electroluminescent diodes. *Applied Physics Letters, 51*(913). doi:10.1063/1.98799.

Taylor, D. (2010, April 9). Motorcyclist charged after recording as an officer pulled him over. *ABC 2 WMAR-TV*. Retrieved from http://www.abc2news.com/dpp/news/crime_checker/ harford_county_crime/motorcyclist-charged-after-recording-as-an-officer-pulled-him-over.

TechBalt. (2008, February 12). Baltimore cop vs skater. Retrieved from http://www.youtube.com/watch?v=1hxOr3q7nrk.

Terry v. Ohio, 392 U.S. 1 (1968).

tesla121. (2008). New video emerges of BAD Baltimore COP. Baltimore, MD: Retrieved from http://www.youtube.com/watch?v=-PnnUJ-OmeM.

The Aging Rebel. (2010, April 9). Getting Got In Maryland. Retrieved from http://www.agingrebel.com/3169.

The History of the Integrated Circuit. (2003, May 5). *Nobelprize.org*. Retrieved from http://nobelprize.org/educational/physics/integrated_circuit/history/.

The New York Times. (2009, March 20). Corrections. A0, 4A. 4. Retrieved from LexisNexis Academic News.

The Voice. (2008). The Voice in the Wilderness [blog]. Retrieved from http://tequilaexpress.blogspot.com/2008/02/great-police-relations.html.

TheDirtyNews. (2009, January 5). Police shooting at BART station - Oscar Grant. *YouTube*. Retrieved from http://www.youtube.com/watch?v=bmJukcFzEX4.

thespecialinterests. (2009, November 15). Bushwick 32 Final 1. *YouTube*. Retrieved from http://www.youtube.com/watch?v=FJ6CP8Wz-N4.

Tocqueville, A. d. (1863). *Democracy in America*. (F. Bowen, Ed.) (2nd ed., Vols. 1-2, Vol. 1). Cambridge: Sever and Francis. I, 265-268.

Tompkins, A. (2010, May 12). Seattle station fails to air video of police beating, firestorm follows. *Poynter*. Retrieved from http://www.poynter.org/latest-news/als-morning-meeting/102599/seattle-station-fails-to-air-video-of-police-beating-firestorm-follows/.

Toshiba. (2004, February 2). *Toshiba camera module includes image sensor, DSP*. Press release. Retrieved from http://www.eetasia.com/ART_8800328859_499501_NP_c5765d2f.HTM.

TRAC. (2004). Under color of law. TRAC Civil Rights Enforcement Reports. Transactional Records Access Clearinghouse. Syracuse, NY: Syracuse University. Retrieved from: http://trac.syr.edu/tracreports/civright/107/.

Trachtenberg, A. (1989). Reading American Photographs: Images as History, Mathew Brady to Walker Evans (1st ed.). New York: Hill and Wang.

Trask, R. B. (1994). Pictures of the Pain: Photography and the Assassination of President Kennedy. Danvers, MA: Yeoman Press.

Tucker, E. (2011, September 20). 2 officers indicted in Md. college student beating. Associated Press. Retrieved from http://www.wtop.com/?nid=539&sid=2554642.

Tucker, J. (2009, January 3). No cuffs on man shot by officer, BART maintains. *San Francisco Chronicle*. Retrieved from http://www.sfgate.com/cgi-bin/article.cgi?f=/c/a/2009/01/02/BAL91534TE.DTL.

Tucker, J., Zito, K., and Knight, H. (2009, January 1). Deadly BART brawl - officer shoots rider, 22. *San Francisco Chronicle*. Retrieved from http://www.sfgate .com/cgi-bin/article.cgi?f=/c/a/2009/01/01/BAB9152I2Q.DTL&tsp=1.

Twyman, M. (1998). *The British Library Guide to Printing : History and Techniques*. British Library guides. Toronto: University of Toronto Press.

U.S. Code. (1871). Civil action for deprivation of rights. Title 42 Chapter 21 Subchapter I § 1983. Retrieved from: http://www.law.cornell.edu/uscode/42/1983.html.

U.S. Code. (n.d.). Conspiracy against rights. Title 18 Part I Chapter 13 § 241. Retrieved from: http://www.law.cornell.edu/uscode/18/241.html.

U.S. Code. (n.d.). Deprivation of rights under color of law. Title 18 Part I Chapter 13 § 242. Retrieved from: http://www.law.cornell.edu/uscode/18/242.html.

U.S. Code. (n.d.). Part B—Police Pattern or Practice: Cause of action. Title 42 Chapter 136 Subchapter IX § 14141. Retrieved from: http://www.law.cornell.edu/ uscode/html/ uscode42/usc_sec_42_00014141----000-.html.

U.S. Department of Justice. (2011a, December 16). Investigation of the Seattle Police Department. Civil Rights Division. Retrieved from http://www.justice.gov/crt/about/spl/ seattlepd.php.

U.S. Department of Justice. (2011b, December 16). Findings letter. Civil Rights Division. Retrieved from http://www.justice.gov/crt/about/spl/documents/spd_findletter_12-16-11.pdf.

U.S. Department of Justice. (2011c, December 16). Justice Department releases investigative findings on the Seattle Police Department. Press release. Office of Public Affairs. Retrieved from http://www.king5.com/home/related/DOJ-press-release-on-Seattle-PD-civil-rights-investigation-135742463.html.

U.S. House of Representatives. (1991). Report No. 102-242, 137-138.

Uhle, F. (2005). *Recreation: Leisure in American Society*. Detroit, MI: Thomson/Gale.

Underwood, M. (2008, February 12). Baltimore Officer Suspended After Video Surfaces of Him Berating Skateboarder. Fox News. Retrieved from http://www.foxnews.com/story/ 0,2933,330501,00.html.

United States Department of Justice. (2012, January 10). Statement of interest of the United States. Retrieved from http://ia700804.us.archive.org/32/items/ gov.uscourts.mdd.194954/ gov.uscourts.mdd.194954.24.0.pdf.

United States. National Advisory Commission on Civil Disorders. (1968). *Report of the National Advisory Commission on Civil Disorders.* Washington: For sale by the Supt. of Docs., U.S. Government Printing Office.

United States. Wickersham Commission, Wickersham, G. W., Monroe, D. G., Garrett, E. W., & Vollmer, A. (1931). *Report on the Police.* Reprint edition (1971). *Police in America.* New York, NY: Arno Press and The New York Times.

United States v. City of Seattle, Case No. 2:12-cv-01282-JLR. (2012, September 21). Stipulation and Order for Modification and for Entry of Preliminary Approval of the Parties' Settlement Agreement and Stipulated Order of Resolution. Retrieved from http://www.justice.gov/crt/about/spl/documents/spd_orderapprovingsettlement_9-21-12.pdf.

Uscanga, D. (November 2010). aTube catcher studio suite (version 2.2.520) [computer program]. Retrieved from http://atube-catcher.dsnetwb.com.

Vacche, A. D. (2002). *The Visual Turn: Classical Film Theory and Art History.* Piscataway, NJ: Rutgers University Press.

Valcourt, D. (2012, October 22). Prince George's County Will Pay $3.6M Settlement In College Park Police Assault Case. *CBS Baltimore*. Retrieved from http://baltimore.cbslocal.com/2012/10/ 22/pg-county-will-pay-3-6m-settlement-in-college-park-police-assault-case/

Van Tassell, R. G. (2013). Walking a Thin Blue Line: Balancing the Citizen's Right to Record Police Officers Against Officer Privacy. *Brigham Young University Law Review*, 2013(1), 183–212.

Vick, C. P. (2007, April 24). KH-11 KENNAN reconnaissance imaging spacecraft. *GlobalSecurity.org*. Retrieved from http://www.globalsecurity.org/space/systems/kh-11.htm.

Vila, B., & Morris, C. (1999). *The Role of Police in American Society: A Documentary History*. Primary documents in American history and contemporary issues. Westport, CT: Greenwood Press.

Volokh, E. (2008, January 30). The dark side of "privacy protections," continued. *The Volokh Conspiracy*. Retrieved from http://volokh.com/archives/ archive_2008_01_27-2008_02_02.shtml#1201752162.

Wacom. (2011). Wacom Company History. Retrieved from http://www.wacom.co.jp/ corporate/en/history.html.

Wagner, P. (2010, December 3). FBI questioning Prince George's Co. Police Officers about College Park student beating. *myFOXdc.com*. Retrieved from http://www.myfoxdc.com/ dpp/news/maryland/fbi-questioning-prince-georges-co-police-officers-about-college-park-riot-120210.

Walker, D., & United States. National Commission on the Causes and Prevention of Violence. (1968). *Rights in Conflict: Convention Week in Chicago, August, 25-29, 1968; A Report*. New York: Dutton.

Walker, S. (1977). A Critical History of Police Reform: The Emergence of Professionalism. Lexington, Mass. Lexington Books.

Walling, G. W. (1887). Recollections of a New York Chief of Police: An Official Record of Thirty-Eight Years as Patrolman, Detective, Captain, Inspector and Chief of the New York Police. New York: Caxton Book Concern.

Wallman, B. (2007a, December 23). Fort Lauderdale police officer cleared in kick case: Video captured Lauderdale incident. Fort Lauderdale, FL: *South Florida Sun-Sentinel*. Retrieved from ProQuest Newsstand.

Wallman, B. (2007b, December 23). Fort Lauderdale police officer cleared in kick case: Video captured Lauderdale incident. Washington, DC: *McClatchy - Tribune Business News*. Retrieved from ProQuest Newsstand.

Wallman, B. (2009, April 22). Police may lose dash-cams. Fort Lauderdale, FL: *South Florida Sun-Sentinel*. Retrieved from http://articles.sun-sentinel.com/2009-04-22/news/ 0904220062_1_video-cameras-patrol-cars-swing-vote.

Walterlx. (2007, December 23). Fort Lauderdale police officer cleared in kick case. CubaNews Yahoo Group. Retrieved from http://groups.yahoo.com/group/CubaNews/message/77240.

Warren, S. D., & Brandeis, L. D. (1890). The right to privacy. *Harvard Law Review, 4*, 193-220.

Washington Market. (ca. 1880). [albumen print]. (X2010.11.3671, MNY217942). Museum of the City of New York. 1220 Fifth Avenue, New York, NY. Retrieved from http://collections.mcny.org.

Washington v. Flora, 845 P.l2d 1355, 1358 (Wash. 1992).

Washingtonian. (2005, August). Suspect is armed with TV camera! Capital Comment. *Washington Magazine*. 9.

Wasserman, H. M. (2009). Orwell's vision: Video and the future of civil rights enforcement. *Maryland Law Review (68)*. 600. Retrieved from LexisNexis Academic.

Waterman, T. (2011, September 23). Boston court ruling affirms citizens' right to record officials. *Greater Boston*. WGBH. Retrieved from http://www.wgbh.org/articles/Boston-Court-Ruling-Affirms-Citizens-Right-To-Record-Officials-4342.

WBAL & Lang, R. (2010a, August 26). Father of teen in Rivieri video speaks out. Baltimore, MD: WBAL Radio. Retrieved from http://www.wbal.com/absolutenm/templates/ story.aspx?articleid=57479&zoneid=3.

WBAL & Lang, R. (2010b, August 28). WBAL exclusive: Officer Rivieri speaks out on firing. Baltimore, MD: WBAL Radio. Retrieved from http://www.wbal.com/absolutenm/templates/story.aspx?articleid=57608& zoneid=3.

WBAL & Patti, J. (2010, July 20). Trial board clears Officer Rivieri. Baltimore, MD: WBAL Radio. Retrieved from http://www.wbal.com/absolutenm/templates/story.aspx? articleid=54995&zoneid=3.

WBAL. (2008, February 12). Cop suspended after skateboarder confrontation. Baltimore, MD: WBAL Radio. Retrieved from http://wbal.com/stories/templates/news.aspx? articleid=1928&zoneid=2.

WBAL. (2010a, August 26). Union angry at officer's firing. Baltimore, MD: WBAL TV. Retrieved from http://www.youtube.com/watch?v=xpK9B3tlir0 and from http://www.wbal.com/absolutenm/articlefiles/57408-Bob%20Cherry%20reacts%20to%20officer%20firing.MP3.

WBAL. (2010b, August 27). Fallout continues after Officer Rivieri firing. Baltimore, MD: WBAL TV. Retrieved from http://www.youtube.com/watch?v=afAVDwCyiZ4.

WBAL. (2010c, August 30). Rivieri: There was more to the video. Baltimore, MD: WBAL TV. Retrieved from http://www.youtube.com/watch?v=F8pEaHRCfEU.

WBAL. (2010d, August 30). Mayor Rawlings-Blake: "Rehire Officer Rivieri NOW!". C4 Commentary. Baltimore, MD: WBAL Radio. Retrieved from http://www.wbal.com/absolutenm/templates/story.aspx?articleid=57669&zoneid=12.

WCBS. (2008, September 24). NYPD Taser death sparks outrage in Bed-Stuy. Retrieved from http://wcbstv.com/local/taser.death.nypd.2.824825.html.

WCBS-TV. (2008a, October 26). CBS2 photographer arrested. Closed captioned. *CBS 2 HD News*. Retrieved from http://newyork.cbslocal.com/2008/10/26/.

WCBS-TV. (2008b, October 27). Cop suspended for chokehold on CBS 2 cameraman. *WCBSTV.com*. Retrieved from http://wcbstv.com/topstories/cbs.photographer.arrested.2. 849897.html.

Webb, R. C. (2005). Tele-visionaries: The People Behind the Invention of Television. Hoboken, N.J.: Wiley-Interscience.

Weegee. (1945). *Naked City*. New York: Essential Books.

Weeks v. United States, 232 U.S. 383 (1914).

Weinblatt, R. (2008, February 14). Police Trainer News Interview Response Cop vs. Skater. Orlando, FL: WOFL-TV Fox 35 News. Retrieved from http://www.youtube.com/ watch?v=-g0bfWgRFMk.

Weiser, B. (2008, May 12). Police in gun searches face disbelief in court. *The New York Times*.

Weiss, M., Cornell, K., and Murphy, K. (2008, July 27). Video: Cop body-checks bicyclist. *New York Post*. Retrieved from http://www.nypost.com/p/news/regional/ video_cop_body_checks_bicyclist_c3oCukbdXAmw9wzbNFOwTL.

West, N. M. (2000). *Kodak and the Lens of Nostalgia. Cultural Frames, Framing Culture*. Charlottesville: University Press of Virginia.

Westerbeck, C., & Meyerowitz, J. (1994). *Bystander: A History of Street Photography* (1st ed.). Boston: Little, Brown and Co.

Westley, W. A. (1970). Violence and the Police: A Sociological Study Of Law, Custom, and Morality. Cambridge, MA: MIT.

White, D. (1980). Prime time-anytime. *Saturday Evening Post, 252*(6), 82-97.

White, E. J., & American Civil Liberties Union. (n.d.). What to do if you're stopped by police. *ACLU.org*. Retrieved from http://www.aclu.org/immigrants-rights-racial-justice/elon-james-white-what-do-if-youre-stopped-police.

White, J. H. (1896, May 11). *Herald Square*. Edison Manufacturing Co. [1 film reel of 1 (ca. 1 min., ca. 16 ft.) : si., b&w ; 35 mm. viewing print]. Library of Congress: American Memory (FEC 8061). Library of Congress Motion Picture, Broadcasting and Recorded Sound Division, Washington, D. C. Retrieved from: http://memory.loc.gov/ mbrs/edmp/4039.mpg.

Wiesenfeld, L. P. (Ed.). (2001). *The World Almanac and Book of Facts, 2001*. Mahway, NJ: World Almanac Books.

Willis, N. P., Porter, T. O., & Talbot, H. F. (1839 4). The Pencil of Nature: A New Discovery. *The Corsair*, *1*(5), 70-72.

Wilson, C. P. (2000). Cop Knowledge: Police Power and Cultural Narrative in Twentieth Century America. Chicago: Univ. of Chicago Press.

Wilson, E. L. (1900, September). Copyright and the American photographer. *Wilson's Photographic Magazine 37*(525) 385-386.

Wilson, F. H. (1890). Some Comments on the German Group. *Wilson's Photographic Magazine* (aka *Photographic Journal of America), XXVII*, 345-347.

Wilson, R. (Ed.). (2007). Scope: Looking back: To the beginnings of handheld videocameras. *Electrical Design News*, *52*(6), 92.

Wimmer, K., & Blevins, J. (2005 8). Rights of Journalists on Public Streets. Memorandum, . Retrieved from http://www.nppa.org/news_and_events/news/2005/08/rights.pdf.

Wimmer, R. D., & Dominick, J. R. (2000). *Mass Media Research: An Introduction.* Wadsworth Series in Mass Communication and Journalism (6th ed.). Belmont, CA: Wadsworth.

Winston, A. (2011, June 13). Mehserle released after 11 months in LA county jail – now what? *KALW*. Retrieved from http://blog.sfgate.com/kalw/2011/06/13/mehserle-released-after-11-months-in-la-county-jail-now-what/.

Winter, B. (frozenphoenixprod). (2010, March 4). College Park riot March 4 2010. *YouTube*. Retrieved from http://www.youtube.com/watch?v=1o9krzgiV5A.

Wiredpig. (2011, May 3). Police technology roundup for April 30 – May 3. *WPS Police Tech - A Police Technology Blog*. Retrieved from http://tech.wiredpig.us/2011/05/ police-technology-roundup-for-april-30-may-3/.

WJLA. (2005, May 13). DC news 7 reporter gets the smackdown by cops. *YouTube*. Retrieved from http://www.wtop.com/?nid=&sid=518966.

WJLA. (2010a, March 03). Charging documents for John J. McKenna. Retrieved from http://static.wjla.com/documents/pgpdcharging041210x.pdf.

WJLA. (2010b, April 8). Maryland State Police press conference. Retrieved from http://www.wjla.com/news/stories/0410/725740.html.

WJZ. (2008, February 11). Officer Suspended After Appearing In YouTube Video. Retrieved from http://wjz.com/local/youtube.salvatore.rivieri.2.651552.html.

WMAR. (2008, February 15). New charges against city cop. Baltimore, MD: WMAR ABC2News.com. Retrieved from http://www.abc2news.com/mostpopular/ story.aspx?content_id=4f8b7c0f-6d4a-47b7-9e9b-569803ea7db8.

WMAR. (2010, August 25). Police officer fired. Baltimore, MD: WMAR ABC 2 News. Retrieved from http://www.youtube.com/watch?v=x9Z5znRVZxg.

WMAR-TV. (2010a, April 8). Brian Kuebler interviews Anthony Graber. Retrieved from http://www.youtube.com/watch?v=QNcDGqzAB30.

WMAR-TV. (2010b, April 8). Greg Shipley at Maryland State Police press conference. Retrieved from http://www.youtube.com/watch?v=QNcDGqzAB30.

WNBC. (2008, September 24). Man falls to his death after being Tasered by police. Developing story. Retrieved from http://www.wnbc.com/news/17550915/detail.html?dl=mainclick.

Wolff, N. (2007). *The Professional Photographers Legal Handbook.* New York: Allworth Press.

WTOP. (2005, May 13). Prince George's officials eye police cameras. *WTOP Radio.* Retrieved from http://www.wtopnews.com/?nid=708&sid=1577281.

WTOP (WTOPradio). (2010a, April 13). Eyewitness video of Maryland Beating. *YouTube*. Retrieved from http://www.youtube.com/watch?v=08OR_Dr1l8s.

WTOP (WTOPradio). (2010b, April 14). Eyewitness video of police beating UMD student. *YouTube*. Retrieved from http://www.youtube.com/watch?v=ajR6Fga8tsw.

WTOP. (2010c, April 21). Missing tape in UMd. beating surfaces. *WTOP.com*. Retrieved from http://www.wtop.com/?nid=&sid=1938732.

Yaniv, O. (2010, July 15). No jail time for biker-slamming former officer. *Daily News*. Sports final edition. News section. 27. New York, NY. Retrieved from LexisNexis Academic News.

Yin, R. K. (2003a). *Case Study Research: Design and Methods*. Applied social research methods series (3rd ed., Vol. 5). Thousand Oaks, CA: Sage Publications.

Yin, R. K. (2003b). *Applications of Case Study Research*. Applied social research methods (2nd ed.). Thousand Oaks, CA: Sage Publications.

YouTube. (2010a). Community guidelines. Retrieved from http://www.youtube.com/t/ community_guidelines.

YouTube. (2010b). Terms of service. Retrieved from http://www.youtube.com/t/terms.

YouTube. (2010c). Account and policies: Content reporting and removals. Retrieved from http://www.google.com/support/youtube/bin/answer.py?hl=en&answer=178909.

YouTube. (2010d). YouTube for mobile. Retrieved from: http://www.youtube.com/mobile.

YouTube. (2011a). Community Guidelines. Retrieved from http://www.youtube.com/t/ community_guidelines.

YouTube. (2011b). DMCA policy. Retrieved from: http://www.youtube.com/t/dmca_policy.

YouTube. (2011c, December 30). BART Police shooting of Oscar Grant compiled by YouTube. Retrieved from http://www.youtube.com/topic/GocWH-HkzQU/bart-police-shooting-of-oscar-grant?feature=rhs_rel.

Zapotosky, M. (2012, December 14). Officer who beat U-Md. student is sentenced to home detention. *The Washington Post*. Retrieved from http://www.washingtonpost.com/local/crime/sentencing-scheduled-for-md-office-in-campus-beating/2012/12/14/8ccb0de4-45f5-11e2-8061-253bccfc7532_story.html.

Zapotosky, M., and Castaneda, R. (2010, December 4). Feds now lead probe of police beating. *The Washington Post*. Met 2 Edition. Metro, B06. Retrieved from LexisNexis Academic News.

Zavada, R. J. (1998, October 24). Dissecting the Zapruder Bell & Howell 8mm movie camera. Presentation to the Movie Machine Society Toronto Conference. Retrieved from http://www.jfk-info.com/zavada1.htm.

Zeidman, S. (2013) Whither the Criminal Court: Confronting Stops-and-Frisks. 76 *Alb. L. Rev. 1187* (2012/2013).

Zimmermann, P. R. (1995). *Reel Families: A Social History of Amateur Film*. Bloomington: Indiana Univ. Press.

Case Citations

Balboa Island Village Inn, Inc. v. Lemen, 121 Cal. App. 4th 583 (2004)
Barber v. Time, Inc., 348 Mo. 1199 (1942)
Black Tea Soc. v. City of Boston, 378 F.3d 8, 10 (1st Cir. 2004)
Branzburg v. Hayes 408 U.S. 665 (1972)
Bremmer v. Journal-Tribune Publishing Co., 247 Iowa 817, 829, 76 N.W.2d 762, 769 (1956)
Burnham v. Ianni, 119 F.3d 668, (1997)
Channel 10, Inc. v. Gunnarson, 337 F. Supp. 634 (D. Minn. 1972)
Daily Times Democrat v. Graham, 276 Ala. 380 (1964)
Dayton Newspapers Inc. v. Starick, 345 F.2d 677 (6th Cir. 1965)
Dietemann v. Time, Inc., 449 F.2d 245 (1971)
Fordyce v. City of Seattle, 55 F.3d 436 (9th Cir. 1995)
Forster v. Manchester, 410 Pa. 192 (Pa. 1963)
Gannett Satellite Info. Network, Inc. v. Township of Pennsauken, 709 F. Supp. 530, 536-37 (D.N.J. 1989)
Gazette Publishing Co. v. Cox, Cause No. IP 65-C-528 (S.D. Ind. May 2, 1967)
Gill v. Hearst Publishing Co., 40 Cal. 2d 224; 253 P.2d 441 (1953)
Greer v. Spock, 424 U.S. 828 (1976)
Hague v. CIO, 307 U.S. 496, 59 S. Ct. 954, 83 L. Ed. 1423 (1939)
Heffron v. Int'l Soc. for Krishna Consciousness, Inc., 452 U.S. 640, 650 (1981)
Humiston v. Universal Film Manufacturing Co., 189 App. Div. 467, 472, 178. N.Y. Supp. 752, 756 (lst Dep't 1919)
International News Service v. The Associated Press, 248 U.S. 215 (1918)
Lambert v. Polk County, Iowa, 723 F. Supp. 128, 133 (S.D. Iowa 1989)
Leverton v. Curtis Pub. Co., D. C. Pa., 97 F. Supp. 181, affirmed 192 F.2d 974 (1951)
Moore v. New York Elevated R. Co., 130 N.Y. 523, 29 N. E. 997 (1892)

NY Sess. Laws ch. 132, §§ 1-2 (1903)
Pavesich v. New England Life Ins. Co., 122 Ga. 190 (1905)
Perry v. Los Angeles Police Dep't, 121 F.3d 1365, 1369 (9th Cir. 1997)
Perry Educ. Ass'n v. Perry Local Educators' Ass'n, 460 U.S. 37 (1983)
Porat v. Lincoln Towers Cmty. Ass'n, No. 04Civ.3199, 2005 WL 646093, at 4-5 (S.D.N.Y. March 21, 2005)
Pritchett v. Board of Commissioners of Knox County, 42 Ind. App. 3 (1908)
Roberson v. Rochester Folding Box Co., 171 N.Y. 538 (1902)
Schnell v. City of Chicago, 407 F.2d 1084 (7th Cir. 1969)
Smith v. City of Cummings, 212 F.3d 1332, 1333 (11th Cir. 2000)
Souder v Pendleton Detectives, 88 So. 2d 716 (1956)
Sports and General Press Agency, Ltd., v. "Our Dogs" Publishing Co., Ltd., 2 K.B. 880 (1916)
Thayer v Worcester Post Co., 284 Mass. 160; 187 N.E. 292 (1933)
United for Peace and Justice v. City of New York, 243 F. Supp. 2d 19, 24-25 (S.D.N.Y. 2003)
United States v. Grace, 461 U.S. 171 (1983)
United States v. Kokinda, 497 U.S. 720 (1990)
USCS Const. Amend. 1 (1789)
Wilkinson v. Frost, 832 F.2d 1330 (2d Cir. 1987)

Index

CPSIA information can be obtained at www.ICGtesting.com
Printed in the USA
BVOW03*1330310315

393398BV00001B/1/P

9 781593 327576